Natural Connection

Natural Connection

*What Indigenous
Wisdom & Marginalised
People Teach Us About
Environmental Action*

JOYCELYN LONGDON

SQUARE PEG

13 5 7 9 10 8 6 4 2

Square Peg, an imprint of Vintage, is part of the
Penguin Random House group of companies

Vintage, Penguin Random House UK, One Embassy
Gardens, 8 Viaduct Gardens, London SW11 7BW

penguin.co.uk/vintage
global.penguinrandomhouse.com

First published by Square Peg in 2025

Copyright © Joycelyn Longdon 2025

Joycelyn Longdon has asserted her right to be identified as the author of this
Work in accordance with the Copyright, Designs and Patents Act 1988

Penguin Random House values and supports copyright. Copyright fuels creativity, encourages diverse voices, promotes freedom of expression and supports a vibrant culture. Thank you for purchasing an authorised edition of this book and for respecting intellectual property laws by not reproducing, scanning or distributing any part of it by any means without permission. You are supporting authors and enabling Penguin Random House to continue to publish books for everyone. No part of this book may be used or reproduced in any manner for the purpose of training artificial intelligence technologies or systems. In accordance with Article 4(3) of the DSM Directive 2019/790, Penguin Random House expressly reserves this work from the text and data mining exception.

Typeset in 11.6/ 15.8pt Calluna by Jouve (UK), Milton Keynes
Printed and bound in Great Britain by Clays Ltd, Elcograf S.p.A.

The authorised representative in the EEA is Penguin Random House Ireland,
Morrison Chambers, 32 Nassau Street, Dublin D02 YH68

A CIP catalogue record for this book is available from the British Library

ISBN 9781529902662

Penguin Random House is committed to a sustainable future
for our business, our readers and our planet. This book is made
from Forest Stewardship Council® certified paper.

Contents

Prologue: Taking Root — 1
Introduction: Natural Connection — 3

RAGE

1 Rage Is Resistance — 15
2 The Dawn of the Environmental Justice Movement — 22
3 The Ogoni 9: Nigeria's Fight against Fossil Fuels — 33
4 Following Rivers of Resistance: The Quilombola Fight in Brazil — 41
5 The Original Treehuggers: India's Anti-Deforestation Movement — 48
6 Joy Is the Sister of Rage — 59
7 From Rage to Imagination — 63

IMAGINATION

8 Reclaiming Our Collective Imagination — 67
9 Decolonising Imagination — 78
10 Does a Glacier Mourn Its Death? — 90
11 Parables for the Future — 98
12 From Imagination to Innovation — 108

INNOVATION

13 (How) Will Technology Save the Planet? — 113

CONTENTS

14	Wakandan Cosmology: A Blueprint for Rooted Innovation	126
15	Roots of Innovation	139
16	The Future Lies in the Clay: Iran's Energy-Free Cooling Solutions	147
17	From Innovation to Theory	155

THEORY

18	Theory as Liberation	159
19	It Takes a Lawyer, an Activist and a Storyteller (and You) to Change the World	170
20	Theory as Collective Wisdom	175
21	Straddling Worlds for Resistance and Change	181
22	Systems Change, Not Climate Change	191
23	From Theory to Healing	201

HEALING

24	The Earth Is a Church: Ethiopia's Architecture-Inspired Conservation	205
25	Nature Is a Human Right	213
26	Healing Through the Cracks	228
27	Grief Is the Way to Transformation	238
28	From Grief to Care	248

CARE

29	We Are All We Have	251
30	Beyond the Burden of Climate Care	259
31	Climate Work Is Care Work	267
32	Making Kin with the Earth	278

CONTENTS

Epilogue: Rooted Hope:
Our Natural Connection   290

Acknowledgements   299
Select Bibliography   303
Notes   305
Index   331

*Every forest branch moves differently in the breeze,
but as they sway, they connect at the roots.*

—*Rumi*

Prologue: Taking Root

416 million years ago, a new kind of life began to scramble its way out of the depths of the ocean and the shallows of Earth's wetlands. It slipped, and slimed, and grasped and gasped, dragging itself onto land. Unlike its rootless, leafless predecessors who floated on the surface, this new life broke open the hard, encrusted clay of the Earth and planted itself deep into the soil, creating new subterranean connections. It was during this time, the Devonian period, that rooted life was quietly setting the stage for a revolution that would change everything – a time in our geological history commonly known as the 'age of the fishes', when the ocean covered around 85 per cent of the Earth, and North America, Greenland and Europe were joined as a single land mass.[1] After 30 to 40 million years of taking root on land, settling in, the ancestors of modern trees, such as conifers, palms and horsetails, came into being.[2, 3]

We often perceive roots as grounding, life-giving, nurturing forces. They are sacred and medicinal, essential to life on this planet. The arrival of rooted life on land poured oxygen into the skies, supporting the flourishing of new terrestrial life forms, but also contributing to the mass death of marine life.[4] The late Devonian period saw pulses of mass extinction, with almost 70 per cent of all marine species wiped out.[5] Various theories have sought to explain these extinctions. Some believe glaciation and the subsequent lowering of the global sea level to be the trigger – others point to meteorite impacts.[6] But in two separate studies – from the universities of Cincinnati and Indiana respectively – researchers point to the establishment of tree and plant roots.[7] Their work describes how as the newly formed roots descended into the Earth, they splintered, shattered and fragmented rocks, exposing the previously compacted soil to weathering and erosion and, importantly, triggering the release of nutrients, which were deposited into the world's waterways. The study from Indiana University found spikes of nutrients, predominantly phosphorus, in geochemical records, which catalysed the process of eutrophication, causing algal blooms – the excessive and accelerated growth of algae – that suffocated and killed off vast swathes of marine life.[8] Roots, often imagined as life-bringers, brought both abundance *and* loss in the beginning. It was left up to evolution to nurture new roots and to bring the descendants of this rooted life back into balance in order to regenerate the now depleted phosphorus levels.

As is expected, especially with research looking back in time, opinions are still divided on the true cause of the Devonian extinctions. Regardless, an important lesson emerges for our current moment, one which fills me with hope and motivation. The survival of life on this planet depends on the roots that we nurture and that take hold, the roots we allow to run rampant, and the roots we restrict or redirect. In times of destruction, we must answer the call to intentionally plant new roots of balance and regeneration.

Introduction: Natural Connection

The world as we know it is ending, and a new one prepares to be born. We are living through the transition – a transition that is tumultuous and demanding, inflicting and requiring radical change at the same time. With the knowledge that the word 'radical' is born from *radix*, the Latin word for root, we find that our survival through this radical time, as Patty Krawec from Lac Seul First Nation notes, 'depends on what our roots sink into, wrap around, and bring to the surface'.[1] The destruction we are witnessing unfolding around us can be traced to the violent, oppressive and dying roots of colonialism and late-stage capitalism: to centuries of domination over the land and those most vulnerable or marginalised. Our roots – the cultural, historical, physical and spiritual connections we have with each other and the rest of the living world – have the potential to lead us to pain and suffering, but they can also transport us to the thriving futures we envisage. We must unearth and tend to the roots that will hold us steadfast through the chaos – ideas and ways of living preserved by communities intimately connected to the land and the effects of environmental destruction: ancient, creative, caring and innovative roots that are waiting to be remembered and nurtured. Roots that remind us of our place *in* and relationship *with* the living world.

Often, as a Black woman and environmentalist living in the West, I encounter individuals whose lives have not been touched by environmental disaster or oppression but who have already admitted defeat. Unable to imagine the world outside its current extractive and destructive systems, they are resigned to apathy and inaction. When I compare this with the ingenuity, dynamism and endurance of frontline communities – those facing the worst climate and environmental impacts, creating beautiful movements in spite of bleak, distressing and traumatic pasts and presents – it becomes clearer than ever that the future cannot be built without the experiences and solutions of marginalised and Indigenous communities, supported by a strong coalition of allies. Communities who, though often neglected from mainstream environmental narratives, through their wisdom and suffering have built a 'natural connection' with the living world. Not a racist or paternalistic image of a romantic, docile, idealised connection with nature. Not the imagined harmony of lives spent frolicking through meadows and communing with trees with no technology in sight. Don't get me wrong, I love a good frolic and I try to visit the magical willow that lives in my local nature reserve as often as I can, but possessing a natural connection with the living world is not a connection based on proximity to 'natural spaces' or the ability to subsist completely off the land. What you will find throughout the stories in this book is that the connection Indigenous and marginalised communities have with the living world is raw, messy, emotional and complicated. It is pain and it is duty; it is honour and it is knowing. What emerges through this thicket of love and loss is a connection to the living world that is rooted in the intimate knowledge that we only have each other and the Earth itself. A natural connection is simply the recognition of environmental action as our steadfast commitment to proudly take our place as beings within, not above or outside of, the planet's astoundingly diverse ecosystem. To plant strong roots in our communities and within the Earth: to build better futures. This connection can be found in all places, wherever our hearts and minds are open and receptive. It exists in the city as much as in the country, for everyone, regardless of whether you work in an office or with the soil. The stories shared in these pages show us that no matter how insignificant we feel, our actions matter, and even 'small' movements have the power to irrevocably impact and change our worlds for the better. My hope is that they illuminate the capacity and power that each of you possesses to craft a natural connection with the world too.

This book is the result of my years-long journey exploring and celebrating the diverse roots of environmentalism, and my quest to share these stories with the world, to inspire and bring about positive and lasting change. Growing up in the Western world, holding that identity in tandem with my identity as a member of the Ghanaian diaspora and as a Black woman while working between the UK and Ghana, fundamentally influences way I view the world, our connection to it and to each other. I have witnessed and experienced the tools of oppression, exploitation and domination. I have seen the ways they wreak havoc on our ecosystems and the people who live within them. And I have, with decreasing surprise, seen how they manifest in the ways we interact with environmental action and build movements of change. But I have also had the privilege of witnessing the beauty, and learning the necessity, of dynamism – of openness, of interconnection. Extraction and oppression are not acts unique to the Western world – I have been exposed, albeit much less frequently, to corruption and lack of care in the Global South too – but they lie at the root of Western domination over people and planet. Here, and in all other instances in this book, I follow Milan and Treré's definition of the 'South' where, 'the South is ... not merely a geographical or geopolitical marker but a plural entity subsuming also the different, the underprivileged, the alternative, the resistant and the invisible'.[2]

What the intersecting crises we are now witnessing provide us with is an invaluable opportunity to unearth and resist these acts. Throughout this book, I hope to offer a series of alternatives. My aim is to leave you as transformed as I have been by the histories that precede us, to find a new and grounded view of your relation to the world, and the power you have to reshape it.

I have always been deeply drawn to understanding our place in the world, the impact of the past on our present and the ways in which we can affect the future. My journey started by looking to the skies, studying astrophysics for my first degree. Then, in 2020, I turned to the ground, embarking on a PhD at the University of Cambridge, exploring the role of technology in forest conservation and the resulting impacts on marginalised communities. In the same year, I founded ClimateInColour, an award-winning education platform for the climate curious that makes climate conversations more accessible, diverse and hopeful. In that first year, I researched, wrote and launched ClimateInColour's first educational course, entitled 'The Colonial History of Climate', which explored the interconnectedness of climate science, environmental breakdown and European imperialism. It was a response to the lack of awareness of the colonial origins of the climate crisis, highlighting the ways in which empire and Western philosophy seeded the systems of oppression and extraction at the root of the intersecting environmental and social crises today. From the creation of botanical gardens in India as a strategy to cultivate and profit from crops in unfamiliar climates to pushing out Indigenous knowledge, science and practices of nurturing the land, the colonial era is the prototype for today's environmental and cultural destruction. More than 500 years after the start of this violent process (which most mark as the 1500s), there is, in the twenty-first century, a real hunger for this information.[3] Hundreds of people signed up to take part in these sessions of collective learning, remembering and connection. Together, we traced the paths of climate colonialism all the way to the present day, revealing the ways in which Western worldviews perpetuate and have intensified our disconnection from the living world. Throughout the course, I told the stories that the mainstream environmental movement was too afraid or unwilling to amplify. Stories that, one by one, brush away the dirt that conceals the roots of Western environmentalism: roots that have cultivated a deeply individualistic and exclusionary movement.

Straddling the worlds of activism, academia, conservation, education and the arts, I have connected with a wide range of conscientious, deeply dedicated and creative individuals. People, at various stages of their journeys to environmental action, whose hearts have been torn apart by the collapse they are witnessing, who are in search of community, purpose, faith: in search of roots that will hold them and guide them through this time of collective grief. Many of them have shared with me their pain and fear, their experiences of distance and disconnection in environmental spaces, and it's likely that those of you reading who have never interacted with ClimateInColour before have had similar experiences. You too are stung by the gatekeeping and the expectation, the demand, for self-sacrifice. The lack of tolerance for diversity of thought, identity and lived experience. You twist yourself into awkward and uncomfortable shapes, in the hope that you will be accepted, crowned as sitting on the 'right side' of history. You watch on as the urgency of the climate and environmental crises pushes activists, scientists and campaigners further into disconnected, draining and demanding lifestyles; they are left burnt out, having exhausted their mental and physical resources and capacity for care in their fight for a better world. You have been shocked to learn how forms of Western environmentalism have grown from the same roots as environmental destruction and systems of oppression. You are saddened by the Western ideals of purity and perfectionism that underpinned the eradication of Indigenous peoples from their lands to reinstate the 'pristine' wilderness, and that are seen mirrored in the contemporary practice of militarised, fortress conservation that sees Indigenous and local communities pushed out of their ancestral homes. You are exhausted by the era of late-stage capitalism that saps from you every last drop of energy, and that tries to harden and isolate you. You are witnessing the uprooting of community and the living world at the hands of colonial, capitalistic and monocultural systems: systems that erode our collective resilience in the face of ongoing disaster – systems from which we must become untethered. You are craving community, compassion, freedom and purpose, the ability to hold and be held and feel safe. You may not yet have realised, or want to acknowledge, that you feel these things, but deep down, in our own unique ways, we are all craving a natural connection.

INTRODUCTION: NATURAL CONNECTION

In this book, I will explore six alternative roots we can grow to create a natural connection between ourselves and the living world. These roots represent practices, teachings and considerations for environmental action inspired by the legacies and ongoing resistance of marginalised communities that transcend Western ideas of what we must do or who we must be to effect positive and impactful change. We will explore, together, how rooting in RAGE is a necessary response to systems of oppression but impedes progress without being transformed into action. We will discover how cultivating radical IMAGINATION allows us to resist the logics of colonialism and capitalism that keep us from building visions of more just futures. We will witness how environmental INNOVATION can and must place the wisdom of nature and cultures, past and present, at its core and how THEORY is a way of building collective wisdom to seed collective liberation. The final chapters lead us to new understandings of our connection with the living world and each other, seeing HEALING as a process of recognising our collective grief as an opening for transformation, and cultivating a universal ethic of CARE by making kin. These chapters, the stories held within them and the lessons they offer, show us how we might move towards a world filled with abundance and safety for all beings.

This is not a book of instructions or demands; it is not a catalogue of climate solutions, although there are many, ancient and modern, held within these pages. It is not a toolkit for environmentalists nor a prescription of the roles and actions you must take, as if environmental action were a mere case of ticking off a checklist. This is a book about acknowledging and deeply embodying your existence as a being who is but one strand woven into the infinite tapestry of the living world. It asks how you might see environmentalism not as a hobby, career or chore, not *only* as an act of doing, but also as a way of being in the world. Reconnecting to *why* and *how* you exist rather than *who* you think you should be or *what* you think you should be doing. This book asks you to release yourself from the stereotypes, the guilt, the ego, the exclusion and the burnout, and move towards radical and honest self-expression and discovery. It asks you to treat environmentalism as the big and small, visible and invisible, ways in which you live and act out your deep love and care for others and the planet.

Through the voices, stories and histories of those who have revolutionised the landscape of environmentalism, each chapter uncovers the richness and diversity of approaches to environmental action pioneered by marginalised communities. Many of the stories held within these pages hail not from popular mainstream media, which focus on often white and Western environmentalists. Rather, they amplify the experiences, triumphs and perspectives of the Global Majority – a phrase coined by Rosemary Campbell-Stephens MBE to describe 'people who are Black, Asian, Brown, dual-heritage, indigenous to the global south, and or have been racialised as "ethnic minorities" in the West'.[4] These are the stories of the Afro-Indigenous Brazilian quilombo resisters *finding roots in rage* against environmental degradation of their ancestral lands; of Ken Saro-Wiwa and the Ogoni 9 in Nigeria, who stood up and lost their lives in the fight against environmental destruction by fossil fuel giants like Shell, and the women of Uttar Pradesh's Chamoli district, who mobilised and catalysed a movement of non-violent protests by hugging trees. Stories of the team at Forensic

Architecture *finding roots in theory* to repatriate the cemeteries of enslaved people in the face of petrochemical extraction and destruction, and of Muthoni Masinde, the Kenyan computer scientist using technology and Indigenous knowledge to solve drought issues for rural farmers. We will explore the technologists *finding roots in innovation*, collaborating with Indigenous African communities to build climate-resilient homes from the clay of the Earth, and creatives like Fehinti Balogun *finding roots in imagination*, using theatre to communicate compelling climate stories. We will hear from those *finding roots in healing*, like transdisciplinary artist and educator brontë velez, who is leading the return to love through spiritual and traditional community reconnection; and from Indigenous leaders like Rarámuri scholar Enrique Salmón who find roots in care by fostering a radical intimacy and familiarity with the living world. These stories are complemented by a beautiful constellation of conversations I was fortunate enough to have with writers, philosophers, environmentalists and change-makers who have inspired my own work. These conversations respond to the stories held within each chapter, providing reflections to help us contextualise each root. In the section on RAGE, the Nigerian ecofeminist and climate justice activist Adenike Oladosu teaches us that anger is a natural response to systems of oppression. In IMAGINATION, acclaimed writer Robert Macfarlane teaches us that imagination untethered from Western philosophical thought leads us to a knowing – deep in our bones – that the living world is *alive*. Reflecting on INNOVATION, designer and expert on Indigenous nature-based technologies Julia Watson teaches us that Indigenous knowledge and narratives are essential in building rooted technologies. In the section on THEORY, Miranda Lowe, principal curator and scientist at the Natural History Museum, London, reminds us that theory is not something created by scholars in the ivory tower, but knowledge already held, and lived through, by marginalised communities. From the transcendent philosopher and activist Báyò Akómoláfé, we learn that HEALING is reimagining wounds as portals – openings to new ways of seeing, living and being. In CARE, bestselling author Katherine May makes clear that care is political, and that we must reject the weaponisation of care as an exercising of power. In the epilogue, ROOTED HOPE, esteemed writer Rebecca Solnit teaches us that, in the midst of environmental breakdown, we can feel terrible and yet remain committed, be heartbroken yet know that the future is being made in the present.

This book is an offering and, through it, I hope to rekindle your sense of awe, wonder and connection with the living world, the histories of those who came before us and the stories of those who will shape the world for the better in the future. Take these stories: inhale them, be moved, enraged, excited and energised by them, and let them help you nurture a natural connection with the world.

Rage

1 Rage Is Resistance

Think back to the first time you felt compelled to engage in environmental action, whether that looked like following an environmentalist or activist organisation on social media, joining your company or institution's sustainability team or society, switching your bank to one that doesn't invest in fossil fuels or seeking out local environmental initiatives. What was it that compelled you to act? What emotions were bubbling under the surface? Was it fear, frustration, hope, sadness, guilt, loneliness or helplessness? For many, at the centre of these emotions lies rage. Pure, hot, white rage. In his book *Earth Emotions*, the philosopher Glenn Albrecht explores the range of emotions that arise within us in response to environmental breakdown. Of the seventeen new terms he coined, *terrafurie* describes the shared, extreme anger felt by those who witness the devastation around them and either feel unable to change the direction of that destruction or aim that anger towards challenging the status quo and holding the leaders of the damage to account.[1] In the face of environmental ruin at the hands of extractive, profit-led, consumption-driven practices, and human and non-human loss of life, there seems to be no other rational emotion to feel. The role of anger in social movements is increasingly being researched, with some studies finding 'eco-anger' to be 'uniquely associated with greater engagement in both personal and collective pro-climate behaviours'.[2] Reflecting on her own motivation for action, in an interview with the *Guardian*, youth climate activist Noga Levy-Rapoport, who led the 2019 London climate strikes, noted that the anger she and fellow young climate activists were experiencing was what took them to the streets in the first place – 'we're marching on the streets because we are furious – we are full of rage and terror'.[3] At the time of writing, the most recent study, conducted by Norwegian researchers, found that anger was the most effective emotion in motivating people to get involved with systemic change and highlighted the links between anger and collective action.[4] Another related study observed similar outcomes, finding that experiences or perceptions of injustice that 'provoke group-based anger' catalyse collective rather than individual action.[5] Given that we need to enact radical changes both as individuals and as a society – the dismantling of systems of oppression is just as important as reducing our own personal contributions to the climate crisis – it would seem that the angry activist trope represents one of the most impactful approaches to inspire environmental action.

Often, the rage that activists feel or express is used against them, especially in the media. Used to discount or invalidate the motivations or objectives of their action. Used to brand them as 'dangerous radicals', as acknowledged in the impassioned speech, on the launch of the third IPCC report, of UN Secretary General António Guterres. Anger is an emotion laden with negative connotations, most commonly aggression, blame, hysteria and violence – perceptions rooted in racism and misogyny. In a passionate and sharp essay entitled 'The Case for Climate Rage', the award-winning journalist Amy Westervelt reflects on her own infuriating experiences of being silenced and invalidated by male colleagues as a reaction to her 'emotional' responses to climate breakdown. She observes how, even in storytelling workplaces, where the goal is to communicate more than just the science, where sadness, hope, humour and even alarm are encouraged, climate stories can 'never be too emotional . . . and especially not angry'. Rage and anger are seen to be in complete opposition to intellect and rationality. We shield our eyes from expressions of anger, confronted by and embarrassed on behalf of those – especially women and people of colour, and particularly Black women – who run about us, naked, their hearts all exposed. Instead of matching their bravery, letting their power light fires within our own hearts, we routinely stay quiet, too afraid to let emotion lead. Aristotle once said that passions are 'dangerous roadblocks on the path to becoming fully human'; the anger of the 'masculine' – playing out through war, violence and extraction – is accepted and tolerated, whilst the anger of the 'other', the not-quite-human, is derided and invalidated.[6] At the close of her essay, Westervelt acknowledges how the branding of rage as shameful lies counter to what we know about climate injustice and the heart-wrenching ways the least responsible have become the most impacted; she makes clear that 'the story of climate change, both its history and its future, needs to be told by people who have already experienced injustice and disempowerment, people who are justifiably angry at the way the system works'.

Having engaged in racial justice activism in my teens and early twenties, witnessing the senseless killing of members of my community in the UK and abroad, and experiencing soul-crushing racism and misogyny myself, I felt all too keenly the burning hot coal of injustice, the scorching of deep anger that sits heavy in the stomach. But as a Black woman, often existing in spaces that require a flattening of emotions – where outbursts, expressions of dissatisfaction, critique and frustration are immediately coded as aggressive, where you are expected to sit quietly in the face of oppression – I have often suppressed my feelings of rage. This is not an act unique to me. I come from a long line and broad community of Black women for whom abandoning the self, quieting the growing storm within, is a strategy for survival, physically, mentally and socially. I speak here with the specificity of the Black experience, not to negate the experiences of silencing that most, if not *all*, women around the world face, but to talk about my own community and the distinct challenges they face as women withstanding sexism as well as racism. The foregrounding of the struggle of Black women is an essential part of intersectional feminism, which was pioneered by Black women thinkers such as Angela Davis, Roxane Gay, Audre Lorde, Kimberlé Crenshaw (who coined the phrase), and more recently Bernardine Evaristo. I use the intersectional lens to make space for the nuanced way different forms of oppression are layered on us depending on who we are and where we come from so that our shared experiences don't overshadow the ones we don't all share in the same way. But, for all of us, there's a lot to be mad about.

Like racism and sexism, environmental breakdown is a form of oppression inflicted primarily by the elite and those in the Global North. It's something to be mad about. Accelerating species loss and extinction is something to be mad about. Worsening air quality, especially in the poorest areas, is something to be mad about. Increasingly severe climate disasters, hurricanes, floods, forest fires are things to be mad about. We have a lot to be mad about and as long as our bodies tremble and hearts ache, we must resist the forces that manipulate and control, pushing us further away from that madness. Yet, the response to public demonstrations of rage from environmental activist groups, like Extinction Rebellion and Just Stop Oil, in Britain highlight how divisive rage has become. When activists transform their anger into action, interrupting a popular theatre production and calling out the theatre's ties to fossil fuel destruction, they are taunted and heckled. Vitriol is spewed in their faces for loving the arts so much that they cannot bear to see it covered, metaphorically, in oil. Or when their anger moves them to lie in the road – a tactic that we will later see stems from the dawn of the environmental justice movement, led by Black communities in America – they are ripped apart by the media and, in some instances, physically abused by their fellow citizens.[7] Recent reports have found that 80 per cent of the British public are concerned about climate change, and chances are, most of them feel angry about it. Yet, with the help of the British media, we have been conditioned to characterise anger in response to injustice as unacceptable, and anger in response to rage about injustice as natural. There's a tension here, one that many of us feel and that has been discussed in numerous video essays on social media. As someone who engages with direct protest as part of my personal activism, I understand the frustration of watching people you love, your neighbours, your city, seemingly unphased by destruction and exploitation. Watching people argue over the dinner table, in the comment sections on social media or on the TV about how bad things are and how we are doomed yet refuse to openly perform this rage where it feels it counts – the streets. There also exists an equal level of frustration from people who feel and have witnessed how, at times, being out on the streets is more about telling and showing the world you care – often with a sense of superiority over others – rather than truly calling for transformation. With many of us already so disenfranchised by our social systems and governments, it's hard not to feel that public displays of rage are pointless – that they change nothing. In a world where we are more and more disconnected from our communities and ever more polarised, we take out the rage triggered by intense feelings of lack of agency and power on each other instead of channelling it into transformation. Instead of using our anger to advocate for change, we are trapped in cycles of collective conflict; we forget that without radical action, much bigger disruptions with much more devastating outcomes will befall us, hurting those of us most vulnerable and marginalised in the UK and globally. All the while, those in power continue exploiting us all. What we fail to see, as a collective, is that our rage is a valuable resource. As Audre Lorde says, our 'well-stocked arsenal of anger [is] potentially useful against oppressions ... focused with precision, it can become a powerful source of energy serving progress and change'.[8] In this quote, there are two important words: 'potentially' and 'precision'. Earlier, I reflected on my difficulties in embracing rage in my own environmentalism. It was not rage, the feeling itself, that I was weary of, but rather the consequences of misguided, righteous, shaming, blaming rage. Rage that is disjointed from dreams of harmony and visions of connection. Rage that harms and hurts. Rage that perpetuates the capitalist and colonial behaviours of using, exhausting and devaluing others. Rage is a fire. A fire that can enlighten and warm yet burn and destroy in equal measure. Lorde reminds us of the 'potential' of rage to catalyse positive change, emphasising that this is not a given but something that requires care, thought and intention. We must be 'precise': in the ways we channel our rage, who we inflict it upon and with the visions of liberation that underlie the transformation of that rage into action.

When I trace the marks of anger that embroider my heart, I find that they lead to devastation. My anger comes from a place of loss, a place of despair. Thinking back to one of the early rally cries of Extinction Rebellion – love and rage – it is obvious to me that so much of my anger stems from and is imbued with the deep affection I have for our planet.[9] It is our love for and dismay at the destruction of the living world that must lie at the core of our rage, our *terrafurie*. Writer and activist adrienne maree brown in her article on 'how the wonder of nature can inspire social justice activism', explores the perspectives of different organisers, facilitators and artists who were leading transformation around the world.[10] Whilst reading it, I was struck by the contribution from reproductive rights organiser Jasmine Burnett, who wrote about 'devastation as a source of liberation'. These words encouraged me to reconsider the way I thought about how anger manifests in organising, protest and direct action. These words made clear to me that the rage that boils within us when we are confronted by destruction is the very thing we need to achieve collective liberation.[11] So how do we go about harnessing and focusing our collective rage? How do we reject corrosive, soul-sucking, competitive manifestations of fury and cultivate energising, purpose-holding rage that is not suppressed or left festering in the stomachs of individuals but tended to and transformed by community into action? The answers to these questions lie in the forests, oil fields and towns of communities past and present, fighting and struggling for their non-human kin, the land and themselves. This section is about how rage has been and can be harnessed and transformed to move from a place of destruction to liberation. In the next chapter, we will travel back in time to North Carolina in the 1970s. We will meet the Warren Country community and learn, through their story, how rage – held collectively – shaped actions that would mark the dawn of the environmental justice movement.

2 The Dawn of the Environmental Justice Movement

As a born and bred Londoner, growing up just 5 miles away from Heathrow Airport, I can think of a few examples that show how the transformation of anger into action plays out in city landscapes. London is a city choking on a thick, heavy and persistent cloud of air pollution; a dizzyingly hazardous soup of carbon monoxide, ozone, sulphur dioxide, nitrogen dioxide and particulate matter, the latter two of which exist in quantities so large that they pose a public health emergency, as they cause respiratory and cardiac illnesses, skin irritation, metabolic disorders as well as neurological and reproductive issues.[1,2] This veil of pollution hangs heaviest in the skies over predominantly Black and South Asian communities. Contributing to more than 9,400 deaths a year, air pollution was responsible for the tragic loss of life of nine-year-old Ella Kissi-Debrah from South London, the first person in the UK for whom air pollution was identified as the cause of death.[3] Ella's death, on 15 February 2013, devastated the Black community, who knew all too well that her death was a result of intersecting and compounding injustices: Ella suffered from asthma, a condition that causes 50 per cent of the hospital admissions originating in children from Black and Brown communities.[4] In August 2020, as a response to this enraging reality, three Black and Brown sixth formers from London – Anjali Raman-Middleton, Nyeleti Brauer-Maxaiea and Destiny Boka-Batesa – who attended primary school with Ella, came together to form the charity Choked Up.[5] They were tired of their voices being side-lined and 'outraged that the people most at risk of the health impacts of air pollution were people of colour and working-class communities'.[6] Using artful, creative and direct guerrilla campaigns in heavily affected areas, they turned their anger and devastation into action. In April 2021, they installed fake but realistic 'POLLUTION ZONE' road signs that warned that 'Breathing Kills'. They also demanded that Sadiq Khan, London's mayor, enshrine the right to breathe clean air in UK law through a new Clean Air Act. The campaign, still ongoing, has been a success, with its first three asks being acted upon. This has resulted in

1. a fining policy for red routes, which make up 5 per cent of London's roads but account for 30 per cent of the city's traffic;[7]
2. an expanded Ultra Low Emission Zone (ULEZ), which by the end of 2022, a year after the expansion, saw nitrogen dioxide levels decrease by 46 per cent and carbon monoxide levels reduce by 22 per cent; and
3. an acceleration in the rollout of zero-emission buses with the launch date moved up from 2037 to 2034.[8]

The work of Choked Up is essential, as it directly addresses the systems of oppression that have resulted in redlining in London and that have become commonplace across the UK, US and beyond. Redlining refers to the ongoing process of racial segregation that drives the isolation of marginalised communities into housing projects exposed to poor environmental standards in order to reserve more liveable environments for white and middle-class communities.[9] The roots of the term come from the literal red lines drawn on maps around areas that housed Black communities by government officials. These areas would be completely shaded in red, even if only one Black person was living in the area, and would be marked as low value in terms of house prices.[10] The term has evolved to describe these often covert acts of segregation where communities are marginalised by race, exposing them to lower living standards and quality of life. The legacy of redlining is a long one, but so is the tradition of environmental justice activism. The campaigning by Anjali, Nyeleti and Destiny carries forward the work of generations of activists who resisted harmful segregational policies.

One of the most famous examples of such activism was the PCB resistance in Warren County, North Carolina in the late 1970s. Over three months in the summer of 1978, concealed by the dark of night, a team of men working for the Ward Transformer Company transported an estimated 31,000 gallons of toxic transformer oil along 240 miles of North Carolina roadways.[11] The transformer oil was infused with dangerous chemicals like dioxin, dibenzofurans and, most famously, large quantities of polychlorinated biphenyls (PCBs). As planned by Ward, who had counted on the absorptive nature of the local soil, the chemicals drained into the surrounding lakes, farmland, and even the groundwater across fourteen counties in the state.[12] Just a year prior, the US Environmental Protection Agency had advocated for a ban on the PCBs that were known to cause cancer, in addition to having many other harmful effects.[13] While Ward Transformer Company was held to some level of account, sued by the state of North Carolina for damages, the toxic chemicals remained in the ground wreaking havoc on social and environmental health. Far from serving justice to its people, the state came up with a clean-up plan that would threaten the lives of the most vulnerable. In December 1978, with no prior warning or consultation, the government announced that they would excavate the 211 miles of contaminated soil weighing 50,000 tonnes and dump it in a local farm in the town of Warren County.[14] The town's population was 60 per cent Black and overwhelmingly poor, with many residents having no indoor plumbing.[15] Drawing water from wells located just 10 metres below ground level, many were scarily vulnerable to contamination from toxic pollutants that would seep from the landfill, through the town's highly permeable soil, into the groundwater. Concern spread throughout the community. The health of the people and the state of the environment were in grave danger – something needed to change. With less than two weeks to organise ahead of the public hearing on the issue scheduled for 4 January 1979, the residents of Warren County came together to create the association 'Warren County Citizens Concerned About PCBs'. At the helm of the struggle stood a group of strong and dedicated Black women, including Dollie Burwell, a single mother of two, a devout Christian and a leader at the Coley Springs Missionary Baptist Church. Burwell brought news of the toxic dumping and plans for the landfill, activating and mobilising her fellow congregants at her church.[16] Burwell was already an experienced activist, deeply embedded in the civil rights movement. She was led by her faith, crediting her justice advocacy to the call of God to fight injustices in the name of love. But she felt conflicted about how to go about fighting for that justice, admitting that she did not feel that anger was a good quality to have.[17] It was the words of her reverend – who in a sermon reminded her that 'hope has two daughters, courage and anger' –that went on to drive her action and leadership, making her one of the mothers of the environmental justice movement. Burwell led a community full of rage, a community despairing at their misfortune and the deeply racist actions of corporations and the state who felt them so far from human that poisoning them was not a concern. To those who wielded more power than them, they were 'poor, Black and politically impotent', easily exploited by practices of redlining.[18] Burwell's neighbours were riddled with agonising questions:

> *'Do I belong in a dumping ground?'*
>
> .
> .
> .
>
> *'Am I trash too?'*[19]

Channelling her own anger into love, Burwell, alongside a group of equally committed Black women, nurtured the rage burning deep within their community's stomachs, nourishing it with food. The women cooked warming meals, inviting people to eat and connect at protest meetings in the church.[20] Together, the PCB protesters 'redirected their [collective] anger into a movement', a movement that was as interracial as it was intergenerational.[21]

With this guidance, young and old, Black and white, all joined forks and forces over four years, staging rallies and protests to oppose the plans and stop the landfill from being built. Over that period, Burwell was arrested five times, once, alongside her eight-year-old daughter.[22] Their protest was driven by intersecting issues, that of environmental degradation and racial discrimination – what we would today refer to as environmental injustice. It was clear to the Warren County residents that the health of their environment and of themselves was contingent on the respect and humanity, or lack thereof, offered to them because of their race.

When 1982 came around and construction started on the landfill site, the community ramped up their action. Led by Ken Ferruccio, a local community college teacher, Rev. Luther G. Brown, pastor of Coley Springs Baptist, and Rev. Leon White, field director of the United Church of Christ's Commission for Racial Justice in Raleigh, and inspired by previous civil rights movements, they took to the streets. In October 1982, 125 protesters marched from their homes to the entrance gate of the landfill site to meet the first envoy of trucks transporting tonnes of toxic soil into the newly completed plant. On arrival they were met by sixty North Carolina highway patrol officers in riot gear, outfitted with clubs and guns, a stark contrast to the 'bunch of ragtag protestors, many of whom [were] women and children', surrounded by enormous trucks piled high with mountains of toxic soil.[23] Despite the intimidation, the community stood steadfast in their action. Two leaders of the protest were immediately arrested; the other protestors took to the ground, meeting bodies with tarmac in a show of resistance, blocking the passage of the landfill trucks. That day fifty-five protesters were sent to prison.[24] They were later released on the condition that they would never visit the landfill site again. Defying these attempts at subjugation from the authorities, the community took to the streets in what unfolded as a six-week-long, impassioned and beautiful protest: the PCB Protests of Warren County. Across these six weeks, over 500 community members were sent to jail, including Ben Chavis, one of the leaders of the protest, who was arrested not at one of the protests itself but while driving through the farms and fields of the town.[25] Using the excuse that Chavis was driving too slowly, the officers threw him in jail. Grabbing the bars that stood between him and his freedom, Chavis shouted words that would change the landscape of environmental justice activism and legislation:

'This is racism. This is environmental racism.'[26]

The Warren County protests, to the dismay and despair of the campaigners and church, did not achieve its demand of the closure of the landfill site. However, their action had a lasting impact, heralding the dawn of the environmental justice movement. There had been earlier movements inspired and empowered by the same emotions, such as the creation of the union United Farm Workers of America by Chicano (Mexican) activist César Chávez in 1962. Between the mid 1960s and early 1990s, Chávez brought together Latino farmworkers across the US to fight for their human rights, as well as demand protection from pesticide exposure. Chávez's work left a legacy of sustained, unshakeable and successful non-violent action in the name of justice, but it was the Warren County protests that roused the national consciousness of environmental injustice specifically. Its impact rippled across states, countries and continents to inspire similar resistances to environmental contamination worldwide. The rage within these protesters found resonance in movements that were emerging in parallel with the Warren County Protests throughout the 1980s. These years saw a myriad of communities coming together from around the country. All emerging from locations of enacted or proposed destruction that achieved liberations big and small. One such moment of liberation was the publication of the 'Toxic Waste and Race in the United States Report' by the United Church of Christ in 1987. The report compiled all the stories of destruction and resistance shared by communities impacted by the 400 toxic waste sites across the country. That may not seem groundbreaking at first glance, but it was momentous as this was the first national study on the topic, and data within the report had not been made available for public review until then. The report was followed by relentless campaigning which resulted in activists and collaborators meeting with the Environmental Protection Agency, after which finally, in 1994, Bill Clinton issued an executive order called the 'Federal Actions to Address Environmental Justice in Minority Populations and Low-Income Populations'. This was revolutionary as it was a direct instruction for the government to include environmental justice considerations in decision-making, focusing specifically on addressing concerns of environmental injustices on marginalised communities.[27]

Forty years later, in 2022 and on the anniversary of the building of the Warren County landfill, the tireless and seemingly fruitless work of the PCB protesters, and those who continued their legacy, resulted in the announcement of $3 billion being granted for environmental justice with the creation of the Office of Environmental Justice and External Civil Rights.[28] The announcement was made in Warren County itself by Michael S. Regan, the first Black person to head the Environmental Protection Agency. Reflecting on the Warren County struggle and the massive win of the new office four decades later, Jenny Labalme – a photographer and journalist who documented the entire PCB campaign on the ground – saw their story as something that should give 'people hope in this country and elsewhere in the world that their struggles are not in vain'.[29] And this would turn out to be the case.

Returning to the question we asked in the first chapter about how we might harness our collective rage and transform it through community into action, the Warren County story shows us a way. Creating spaces in which we can congregate, build relationships, eat together and cry together allows us to give a home to our rage. For it to be witnessed, observed and ultimately transformed. This space gives us the energy to demonstrate together, to protect one another. While transforming rage into action doesn't always lead to a 'win' in the obvious sense of the word, it always leads to deeper connection between ourselves, our communities and the Earth. For frontline communities witnessing their health and homes being threatened by environmental destruction, rage is instinctual, it is necessary. But so too is it for us living lives safe from imminent environmental disaster. Just like the residents of Warren County, many of us live in places where water pollution is rife. We live in countries that inflict this environmental destruction on others. Harnessing and transforming rage, as the PCB protesters did, is a refusal to harm the planet and to be harmed by destructive forces ourselves. In doing so, we not only protect ourselves, but also leave behind a legacy that inspires and influences the movements of the future, creating a web of liberation crossing generations.

Often, this web of liberation, woven by marginalised communities of the past and present, is rarely cited when discussing mainstream environmental action. When speaking with Nigerian climate justice activist and ecofeminist Adenike Oladosu, I listened as she shared her deep frustration with Western activist movements that silence and ignore the voices of campaigners from Africa and, more widely, the Global South. She reflected on an interaction with an Australian journalist who had gathered

Oladosu and a group of African activists, asking them in detail about the campaigns they were working on, many of which were affecting their communities. They hoped it would be a chance to amplify their causes and gain essential support for their poorly funded and tireless work. Adenike is a powerful force, fighting fiercely – amongst many things – for the restoration of Lake Chad, once one of the largest bodies of water in Africa that has since 1960 shrunk to merely 10 per cent of its original size.[30] When she read the final published article by the journalist, she was shocked and enraged. The headline read 'Greta Returns', highlighting the Swedish activist's return to the country. 'I was like, wow, like seriously, you just read beautiful things about what we do, and you just covered the old story with Greta and her return; that is really unfair.' Oladosu's frustration lies not with Greta, who has made clear the need for intersectionality in activism, but with the Western media and society that prioritises and validates white activism whilst those most affected are ignored.

By neglecting not only the movements that have come before us but also those that keep alive their legacies in the most affected communities around the world, we miss out on the stories of positive change brought about by channelling rage that can lift us out of malaise and perpetual despair. Stories that give us fresh perspectives, teaching us that while the climate crisis on a global level is overwhelming, there continues to thrive a constellation of movements and campaigns that need our support and energy. Without this outlook, we lose sight of the strings of magic that connect each one of us – and our actions – to so many others, past and present. Also of the knowledge that within us lies the strength of our ancestors and that the ripples created by our actions today will stir inside the change-makers of tomorrow. For me, this knowledge is no better encapsulated than in the legacy of the Ogoni 9 – the revolutionaries who catalysed resistance in Nigeria back in the 1970s against the destruction caused by fossil fuel extraction. In the next chapter, we will explore how decades of campaigning, collaboration, loss and success in Nigeria can and has inspired contemporary resistance against the oil and gas industry.

3 The Ogoni 9: Nigeria's Fight against Fossil Fuels

Nigeria's Niger Delta is Africa's largest wetland and the third largest mangrove forest in the world. Composed of a patchwork of kingdoms, the region is home to over thirty million people deeply connected to and reliant on the land. Enduring colonisation from as early as the 1400s, it was more recently in the 1950s that the region experienced some of the most extreme plunder and devastation – when, in 1956, the Royal Dutch Shell discovered crude oil. This discovery marked the end of the old colonial era of palm oil exploitation and the beginning of six decades of environmental and human suffering. Nigeria is one of the world's biggest exporters of oil, and the fossil fuel has become a key source of economic empowerment for the country, providing 95 per cent of the country's overseas earnings. But as the impact of environmental devastation continues to wreak havoc, it has become clear that, as reported in the *Guardian*, oil is 'more of a curse than a blessing' to the region.[1]

In 1970, the year the first Earth Day was celebrated in the US, the region of Bomu in Ogoniland, Nigeria, witnessed a two-month oil 'blowout'. Everything in a 3-mile radius of the deluge was covered in oil and silt. Ogoniland covers over 1,000 square kilometres of the Southeast district of Rivers State, and the region saw its first two oil wells created in 1958. In response to the 1970 disaster, the leaders of many clans in the region collectively wrote to Shell and the state military governor to make clear their distress and disapproval of the environmental disaster unfolding around them. Despite the effect on agriculture resulting from the destruction of farmland and the impact the spill had on fishing, both key economic and subsistence activities, Shell initially rejected applications for compensation from the community, stating that the spill was caused by third parties during Nigeria's civil war when much damage was done to oil pipelines and infrastructure. Despite this assertion, in 2021 Shell Nigeria paid compensation to settle a lawsuit over the oil spill.[2, 3] It is important to note that just four years earlier, both Shell and BP had been instructed by the British government to pay over £3 million in compensation for a similar disaster that occurred off the coast of England.

According to the international environmental NGO Friends of the Earth International, Ogoniland saw nearly 3,000 separate oil spills between 1976 and 1991, and it was found that life expectancy was ten years lower in the Delta region than in the rest of Nigeria. The oil spill of 1970 radicalised many people in the region, most prominently Ken Saro-Wiwa.

Ken was a supporter of the independence of ethnic states in Nigeria, advocating for the right for states to control the activities that occurred on their land and receive financial benefit from them. This was especially important in Ogoniland where international companies extracted not just oil but immense sums of wealth. According to Shell's own numbers, $5.2 billion was extracted from the region between 1958 and 1993, with the local people seeing no part of it.[4] Ken spent years campaigning and communicating ideas around extraction and colonialism in his community through literature, pamphlets and speeches. He had huge cultural influence and was respected in the region. In 1990, he drafted a bill of rights on behalf of the Ogoni people that was later accepted by the then Nigerian President General Ibrahim Babangid. This was the moment that marked the beginning of the Movement for the Survival of Ogoni People, MOSOP, of which Ken became the vice chairman. The Bill of Rights allowed the Ogoni people to rally behind something tangible and advocate for themselves in political spheres, including the United Nations Sub Commission on Prevention of Discrimination and Protection of Minorities (now known as the United Nations Sub-Commission on the Promotion and Protection of Human Rights). From Greenpeace to the UN, the Ogoni experience, one of oppression and devastation, was catching the eyes of the global community.

Unfortunately, the international interest did nothing to move the Nigerian government to act, leading the members of MOSOP to ramp up their action and fight their case through 'mass action and direct confrontation with the State and the oil companies'.[5] MOSOP had five core demands, which were sent to Shell, Chevron and the Nigerian Petroleum Company in 1992. These demands were:

1. Payment of $6 billion to the community as compensation for the use of the community's land for oil exploration since 1958;
2. Payment of $4 billion in damages specifically pertaining to ecological degradation;
3. Immediate halting of gas flaring and further environmental damage in the region;

4. Immediate enforcement of all oil pipelines to protect against future spills; and
5. Beginning negotiations with the Ogoni people to reach a mutual agreement on future oil extraction.

These requests came with a deadline: The Ogoni people made clear that if the demands were not fulfilled in thirty days, they would take to the streets in mass protest. Their action was met with militarisation. The government banned public meetings, and the military were there to enforce the government's wishes – turning what were to be non-violent protests into bloody clashes. Standing at the helm of these actions, Ken became an easy target for persecution by the government, jailed four times in 1993 alone. But the people were tired – tired of being walked over, exploited and harmed – and soon, MOSOP was joined by a growing number of community-led organisations taking individual and collective action against these injustices.

But the strength and unity of this movement was corroded by one awful event. In 1994, four Ogoni leaders, known to vehemently oppose MOSOP, were found murdered.[6] Ken had previously commented on the dissolving of trust within the community. The day after the murders, despite being nowhere near the crime scene, Ken was arrested.[7] With the leader of the movement gone, the military intensified their approach, cracking down on the widespread activism and subduing the community. Despite the murder being committed by 'a group of youths', Ken and eight other high-ranking men in the activist movement were executed by the Nigerian government. They are remembered as the Ogoni 9. Whilst international human rights organisations like Amnesty International accuse Shell of being complicit in the executions, Shell denies any involvement or connection to the deaths.[8]

In the twenty-three years since the death of Ken and his comrades, Ogoniland and the Niger Delta region continue to pay the price for the world's oil addiction and the extractive violence of companies like Shell. In the years between 2011 and 2022, over 500,000 barrels of oils were spilled in the Niger Delta.[9] The recurring and incessant pollution has irrevocably harmed land and life in the region, with life expectancy now recorded at a shockingly low forty-one years. But, to paraphrase the famous saying, whilst the fossil fuel industry works hard (to kill life on Earth for profit) climate justice activists and campaigners have been working harder, tirelessly fighting for decades to bring justice to the region.

After a series of spills in 2003 and 2004, the people of Goi – a village in Ogoniland – were left with nothing: the land was contaminated, the fish were wiped out and there was no safe place to live in. Despite such setbacks, Ken's spirit of activism has endured, just as he knew and prophesied that it would. The following powerful words were written as part of a speech that Ken was prevented from reading before his execution.[10]

> **'I tell you this, I may be dead, but my ideas will not die.'**

Goi's village chief, Barizaa Dooh, outraged and devastated by the destruction of his community, joined forces with three other local farmers and Friends of the Earth Netherlands to take Shell to court. The case lasted thirteen years, bolstered by continued protest from communities who never stopped fighting to bring justice to their land. In 2021, nearly thirty years after the execution of the Ogoni 9, justice was served. In a ruling from The Hague, Shell Nigeria was found liable for damages in the villages of Goi and Oruma. Shell was also found guilty of damages in three other villages and instructed to pay compensation to each farmer there. Rachel Kennerley, a climate campaigner at Friends of the Earth shared her thoughts on the importance of a case like this: 'For too long, companies like Shell have been shirking their responsibility for the impact of the dirty industry they push on communities around the world. Thirteen years of fighting for justice has finally turned this around, and today's judgement is a wakeup call for polluting companies and governments everywhere'.[11] After the court ruling, Shell Nigeria said it continued to believe the spills were caused by sabotage and in 2022 agreed to pay 15 million euros to communities affected by the oil spills as settlement of the case without admission of liability.[12] For Eric Dooh, Barizaa Dooh's son, the success of the case was a lifted burden: 'Finally, there is some justice for the Nigerian people suffering the consequences of Shell's oil ... this verdict brings hope for the future of the people in the Niger Delta'. It was a bittersweet moment for Dooh, who reflected on the fact that his father, like so many others who had lived in constant struggle and pain, did not live to see the outcome of their action.[13] After decades of struggle and loss, the judgment came as an act of justice for the living as much as the dead. The rage of generations of Ogoni communities had been felt and had resulted in a liberation; though not a liberation for everyone in Ogoniland but one nonetheless of the land and of the people. Rage that has continued to reverberate and inspire modern environmental movements.

On the 21st of May 2024, Shell executives from around the world gathered in London for their annual general meeting (AGM). In the first quarter of the year alone, the company boasted profits of $7.7 billion, and in the same period announced drastic cutbacks to their climate targets, pushing their carbon reduction targets of 15 per cent from

2030 to 2050. In a further blow to climate policy and action, the AGM saw a climate resolution posed by twenty-seven investors with a combined $4 trillion under their management urging Shell to align its targets with the Paris Climate Agreements, including emissions from consumers, rejected by over 80 per cent of shareholders in favour of Shell's own internal 'climate strategy'. Bravely disrupting the dismal events, activists from Fossil Free London sang, cheered and eloquently called out the company's detrimental impact on people and planet. One of those activists was Mikaela Loach: On behalf of We The People Nigeria, a group of community organisers representing the people of the Niger Delta, she warned the Shell board that their days of pillage would soon be over. With tears in her eyes but conviction in her words, she demanded answers on who would be cleaning up the decades-long mess of death and destruction in the Delta region. Shell had only recently announced that they would be divesting from oil extraction in the Niger Delta, which the activists saw as Shell passing on their practice of plunder to a consortium of five other fossil fuel companies, with no detailed plan in place to reinvest in and regenerate the area. It is hard not to feel like, to Shell, the land, as well as its people, is a thing to be exploited and then discarded. It journeys across the world leaving oily tracks of hell in its wake whilst lining its pockets with profits. Yet, Mikaela's warning – that Shell's days of profiteering were numbered – was not an empty threat, but an invocation of the power of rage transformed into liberation.

While stories like that of the Ogoni 9 may on the surface seem isolated, they are connected to a wide ecosystem and network of action from frontline, marginalised and Indigenous communities, who are using their anger and frustration to resist environmental destruction. For the Ogoni the fight was against oil and gas, and for the residents of Warren County, it was against toxic waste.

In the next chapter, we travel to Brazil to learn about the centuries-long struggle of Indigenous Afro-Brazilians who continue resisting oppression and destruction by transforming their rage through community action. Their story brings to the fore an underlying teaching rooted in the previous accounts – that of endurance, of how to not only transform rage into action but sustain it.

4 Following Rivers of Resistance: The Quilombola Fight in Brazil

Deep in the waters of the Ribeira de Iguape (Ribeira River) reigns a powerful queen, revered and feared in equal measure. She is Boiuna, the water serpent; to the Indigenous Brazilian community she is Mãe d'Água (Water Mother). Her haunting song lures fishermen into the depths of the river, leading them to their deaths. But the Ribeira River holds more than just the bodies of those sustaining their communities with the water's bounty. Embedded in the river's banks lies an abundance of gold, which, like Mãe d'Água, seduced Portuguese prospectors and slave owners, those set on exploiting sacred lands and 'unknown peoples', in the seventeenth century. It was around this time that a gold rush began, with 'adventurers' travelling across the Atlantic, making pit stops in West Africa to kidnap and purchase slave labour, poised to make a fortune. Unlike most gold rushes, the Brazilian rush would last for centuries and between 1501 and 1886, over 4.9 million slaves were captured and forced to work so that the prospectors could profit. Little did these so-called adventurers know, as the slave ships landed on the shores of Brazil, a rebellion was already stirring in the hearts of the enslaved.

The Ribeira River runs across the states of São Paulo and Paraná in southeast Brazil and courses through the dense vegetation of Mata Atlântica, the Atlantic Forest. Using this rugged and unpredictable terrain to their advantage, the captured Africans found freedom, escaping deep into the forest along the banks of the river. It wasn't long before the reserves of gold in the region were exhausted, and the mines were abandoned. The slave owners left in the 1790s; with them gone, escaped, previously enslaved Africans established 2,000 settlements in the wider region of the Ribeira Valley. They are now known as the quilombola, part of a larger community of Afro-Brazilians in the São Paulo and Paraná states, who have lived in and safeguarded forests and ecosystems spanning much of the coast of Brazil. The story of the quilombola communities is one of incredible feats of bravery, ingenuity and triumph, and three hundred years later, the quilombolas continue to fight new and dark regimes of destruction.

In the 1950s, the quilombola were affected by continuous displacement as their traditional lands were destroyed. Under the guise of conservation, the Brazilian government began to gazette (mark out) land that overwhelmingly overlapped with that of the quilombolas to establish national parks. With threatening military helicopters patrolling the skies, farmers were evicted, fields were burned, and villages were abandoned. Like many Indigenous and local communities across the world, the quilombolas across Brazil were forced to the fringes of the forests they called home. Now, the traditional lands of only 300 out of the 2,000 quilombo communities are officially acknowledged. (While I have been referring to the Afro-Indigenous Brazilian community as the quilombola, a term that describes Maroons and their descendants, a quilombo is the actual settlement of Maroons).[1]

Marginalised and disconnected from their lands, culture and heritage, it was only in the late 1980s that the Brazilian government awarded the quilombola formal recognition as a community. We must remember that African slaves were still being sold in Brazil as late as the nineteenth century. While this recognition is important from the perspective of identity, it is hard to see any real positive change that has resulted from this formality: Official recognition alone meant nothing in the wake of years of unacknowledged displacement and persecution. By not reckoning with these truths, the Brazilian government avoided true and deep engagement with the communities they had come to harm.

This devastation, while shocking, is unfortunately not anomalous. Brazil has a complex geopolitical history, one that is embroidered with class inequality, colourism, elitism and the persecution of the disenfranchised. Take the 2016 Summer Olympic Games in Rio de Janeiro, for example, an event that was hailed a huge success socially, culturally and infrastructurally. The official event website proudly lists the Games' social credentials: 45 per cent of the competitors were women and it was here that the first ever refugee team participated.[2] Yet, below the surface, lay devastation. Not far from the Olympic Park, the media village was built on a sacred mass quilombo grave, not only severing the community from their ancestors but dispossessing the country of a site of important Afro-Brazilian heritage.[3]

Over 4,000 kilometres north of the Ribeira Valley in Pará is the town of Barcarena. Located within the Amazon, the state is popular with tourists for its jungle treks and tours of the Amazon River. But a stone's throw away from these tourist centres, the community members of Barcarena are fighting for environmental justice. For decades, the town has had to contend with the expansion of the Norwegian company Norsk Hydro, which owns the Amazon's largest aluminium refinery. While Norway continues to be acknowledged as a global leader in sustainability and green infrastructure, beyond the country's borders, in the forests of Brazil, that illusory image starts to fade. The Norsk refinery supplies materials to some of the biggest electric car manufacturers in the world and has been touted as an important driver of Pará's socio-economic development by local governments. But for the quilombo, and the local and Indigenous communities who reside in the area, the refinery is seen as a threat to their entire way of life as well as the health of the people and environment. Along with other heavy industries in the region, the refinery has allegedly polluted many important rivers, filling up these ancient waterways with waste lead, aluminium and cadmium. Whilst Norsk Hydro denies that its plant has had any negative health implications, it is feared that the pollution has destroyed soil productivity in the areas surrounding the industrial plants and the children no longer dare to play in the *igarapés*, small streams that run off the Amazon River. Leading the fight against this painful environmental destruction is the quilombo and Indigenous activist Maria do Socorro Silva, whose roots to action can be traced into the forest where she grew up. In an interview with the magazine *Atmos*, Silva bravely recounts atrocious acts of sexual violence she and other young women in her community endured from 'gringos' – white men who had entered their community.[4] Silva 'saw herself melded with the forest', realising that the violence enacted on her was reflected in the destruction and extraction taking place within the forest. Now in her late fifties, Silva is a matriarch in her community, sitting at the helm of a 20,000-strong group of activists and advocates: She leads direct action and protests against the degradation of their lands and advocates for her community in policy-making spheres.

This work, of frontline environmental defence, is

deadly. Those who stand up against the state often face fatal outcomes, as did community, environmental and human rights activist and city councillor Marielle Franco, who was a leading voice against violence against Indigenous communities and inequality in Brazil. On 14 March 2018, Marielle was killed in a targeted shooting. Two of the suspects were politicians.[5] Under the rule of the former right-wing president Jair Bolsanaro, who led the country from 2019 to 2022, the Brazilian government conducted a ruthless and sustained campaign against environmental defenders, including the Afro-descendant quilombola.[6] According to Global Witness, an organisation that advocates for environmental justice at the intersection of natural resource exploitation, conflict, poverty, corruption and human rights abuses, nearly 75 per cent of all reported attacks on land defenders in Brazil occur in the regions around the Amazon. In 2018, Pará state saw the most number of environmental murders in Brazil, many of which are thought to have been state-sponsored.[7] Yet, despite the dangers of land defence, quilombo communities continue to fight.

The threat of death is one land defenders like Silva know all too well: Silva had to protect her home with wrought iron bars when she was met with violent threats and break-ins. And death threats have not always come through direct forms of violence for Silva: Both she and her husband suffered from cancer, which they believe stemmed from the toxic pollution in their area. Despite these challenges, Silva's drive is enduring. She has continued to work with communities from the Pará region, organising protests and rallies, drawing the attention of large publications to bring her community's struggle to light. In an interview with the *Guardian*, she made her stance clear:

> **'Will we fight this? Yes. Will more die? Yes!**
> **They kill the water, the air, and the animals.**
> **They should be put in prison.'[8]**

Reflecting on her account, my heart felt heavy as I thought of what the quilombo communities continue to suffer. But, I am also in awe, humbled to witness the raw and unending love Maria and her community have for their land and their fervent action to protect it. Their connection to the forest and to each other teaches us essential lessons about what it means to sustain environmental action. What it means to have the stamina to carry forward the legacies of our ancestors and fight against ongoing destruction regardless of outcome. The quilombo teach us that the ability to hold and harness rage well can be an heirloom, passed through generations for the continued protection of the Earth and its inhabitants. A lesson that urges us to reflect on not only what we materially leave for future generations, but also what we leave them culturally and spiritually. Through our actions, do we leave them lessons of impatience, apathy and disconnection or instead those of determination and perseverance? The quilombo story shows us that rage is the fuel needed not only to ignite action but also to sustain it in the darkest of times. Their struggle is not straightforward, not a linear route to liberation; there are as many heartbreaking losses as glimmers of hope, and still the quilombo press on, letting their rage power their resistance in the long run.

As with many phenomena on this planet, liberation is

not always reached through a linear path; consider that many march routes are cyclical, starting and ending in the same place. Sometimes liberation through direct action is not achieved by marching at all but by building a path for others to follow.

And, as we'll see in the next chapter, sometimes, all it takes is a hug.

5 The Original Treehuggers: India's Anti-Deforestation Movement

A phrase notoriously associated with movements of non-violent action, including environmental action, is 'tree-hugger'. This phrase originated from the translation of the Hindi word *chipko*, which means *to cling, to hug* or *to embrace*.[1] The word would inspire and become synonymous with the Chipko Movement that emerged in the 1970s to protect sacred and beloved forests across the Indian state of Uttar Pradesh – the centre of which lies eight hours from New Delhi – and beyond. According to an analysis conducted by Vandana Shiva, a pioneering ecofeminist, scholar and activist, in the late 1980s, the seeds of the Chipko Movement were planted many decades earlier, in the 1930s. When the British colonialists came to power in India, they introduced drastic changes in forest management and use. These new methods centred on the privatisation and subsequent exploitation of the important and necessary forest ecosystems and stood in stark contrast to traditional community-managed forestry techniques that previously conserved the commons. This continued unchecked for a long time until in 1930 and 1931, villagers from Madhya Pradesh, another Indian state, came together in peaceful protest known as 'Forest Satyagraha' to win back their land and halt the ongoing environmental destruction in forests bordering them.[2] 'Forest Satyagraha' was based on 'non-cooperative' and non-violent protest, considered to be the 'the most important political weapon' in India, and was influenced by the methods and theories of Mahatma Gandhi.

Before we continue this story, it is important to note that while Mahatma Gandhi is revered and respected as a peacebuilder and leader, we must acknowledge that he was also controversial as an activist because of his surprisingly racist attitudes towards Black people. These beliefs were made clear during his time living in South Africa, where he advocated for the British colonial administration to acknowledge the superiority of the Indian population to the local African communities, writing: 'A general belief seems to prevail in the Colony that the Indians are little better, if at all, than savages or the Natives of Africa. Even the children are taught to believe in that manner, with the result that the Indian is being dragged down to the position of a raw Kaffir.'[3] Nevertheless, his activism cemented non-violent protest as a key method of resistance in India, from the Himalayas to the Western Ghats. The seeds of resistance had been sown and, over the following years, would lead to a string of some of the most famous peaceful environmental protests in history resisting commercial logging and deforestation in the region.

Forty years after the Forest Satyagraha in Madhya Pradesh, this form of peaceful action was again used in an environmental movement in the Alaknanda Valley in Uttarakhand. It started with the Dasholi Gram Swarajya Mandal (DGSM), a worker's co-operative headed by Indian Gandhian environmentalist and social activist Chandi Prasad Bhatt. In 1970, the region was shattered by a disastrous flood, an event that washed away 1,000 square kilometres of land – about four fifths the size of the city of Rome.[4] The event would make the local residents aware of the link between deforestation, landslides and floods.[5] Throughout the early 1970s, the DGSM led peaceful protests and educated their fellow villagers on the relationship between local prosperity, or the lack of it, and the harmful policies of the forest department of the Indian government. In 1973 the forest department granted permission to Simon Company, a manufacturer of sporting goods and equipment, to fell 2,500 trees in Mandal forest in the region.[6] The news brought out people in droves and they marched boldly, joyously and determinedly to prevent the felling of the trees, playing traditional instruments and singing folk songs, finally emerging victorious in their action.[7] But this would not be the villagers' last confrontation with loggers in the region as further threats to the forest were arising. In the face of these mounting challenges, Bhatt needed a new plan, a new tactic, to underpin the already strong resistance unfolding in the hills. Alongside Sunderlal Bahuguna, one of India's best-known environmentalists, Bhatt called out to all those living in the Himalayas, men and women alike. Bahuguna and Bhatt asked of their people one simple act – to join hands, embrace the trees and stop the loggers from cutting them down. Dozens of people hugging trees across the mountains – it was a powerful symbol, one that declared,

'Our bodies before our trees'

and laid the foundations for what would become the Chipko Movement.[8]

By 1974 the forest department set in motion a more targeted and divisive plan to gain back control. Bhatt was restricted from protesting on a day marked for felling in the Reni forest; on the same day, the forest department drew away all the men from the local village under the pretence of providing compensation for their lands.[9, 10] The responsibility of saving the trees was left in the hands of the women of the village, under the fearless leadership of Gaura Devi, who would later be called the 'Mother of the Chipko Movement'. Widowed at the age of twenty-two, Devi was keenly aware of the experiences of loss and the domestic burden she and the women around her carried. She and her community were angry at the impact of logging on the women's ability to feed, serve and provide for their families; they were angry at the dwindling number of local land-based jobs, which meant that their husbands had to go away from them to find work in larger cities. It was in this rage that Devi rooted her call to protect and defend the community's forests. Devi was incredibly spiritual, referring to the forests as her gods. From women's rights to environmental protection, Devi was acutely attuned to the struggles of her people, dedicating her time to these causes, trusted and respected by the women in the community. With this trust, she managed to assemble twenty-seven ready and willing women alongside whom she planned to confront the loggers who had unexpectedly arrived in their village, hungry to fell their sacred trees.[11] The women marched into the forest and right up to the logging contractors, making it clear that they would rather be shot than give them access to the trees. Gaura Devi's proclamations were met with verbal abuse and intimidation with guns. Still, the women persisted. Remembering the suggestions of Bhatt, they joined hands and hugged the trees, refusing to move. The contractors were shocked by this show of defiance and eventually left the site after four days. With the trees still standing, stories of the women's triumph travelled far and wide across the region.

The action of the women that day resulted in real change in policy, with the state government issuing a ten-year ban on commercial logging in the area. Recounting their actions in an interview with Indian historian and environmentalist Ramachandra Guha, Devi described how their action that day 'was not a question of planned organisation ... rather it happened spontaneously. Our men were out of the village, so we had to come forward and protect the trees'.[12] Not only did the Chipko Movement set a new precedent for successful non-violent environmental action, but it also challenged deep-set beliefs about the role of gender in activism, paving the way for women's participation and leadership in Indian environmental engagement.

We have seen this legacy find expression most recently in the history-making farmers' protests in India in 2020–2021 that saw hundreds of thousands of farmers come together to demand the repeal of three livelihood-threatening agricultural laws enacted by the Indian government. If instated, the laws would allow the privatisation of the country's agricultural markets and the abolishing of the minimum support price, a price set by the government to ensure that farmers received a minimum income from essential crops. Reaching its zenith in 2021, the protests, although portrayed in the media as a predominantly male movement, were deeply imbued with the power of India's female farmers. In January of 2021, the Chief Justice of India ordered that women, alongside the elderly, return home from the protests.[13] But voicing the indignation of the protesting women Jasbir Kaur, a seventy-four-year-old farmer from western Uttar Pradesh, boldly stated, 'Why should we go back? This is not just the men's protest. We toil in the fields alongside the men. Who are we—if not farmers?'[14] While 85 per cent of women in rural India contribute to the country's agricultural output, their work remains largely invisible. Land ownership for women farmers sits at a measly 13 per cent and the privatisation of the agricultural industry would only further threaten their already precarious livelihoods.[15] The farmers' protests provided a platform for these women to take centre stage, to make their voices heard and take their positions as individuals on the frontlines of injustice.

The Chipko Movement and the non-violent form of protest it popularised are important examples of the many routes to liberation that can stem from the same source of motivation. There is a simple but powerful magic that emerges from the act of planting oneself in the company of others to resist injustice. It is this energy that is woven into the stories of action we have explored in this section. From the 1970s to the present day, from Brazil to the UK, rage has acted as a conduit for the formation of community, connection and resistance. In all these stories, we see the coming together of communities, bound by struggle, exploitation, extraction and, undeniably, anger. Rage is often discounted as an emotional drive and termed unstable and irrational. But that was the emotion which swelled under the seemingly calm exterior of the women of the Chipko Movement who so bravely protected their forests. The lack of violence or outward fury did not negate the rage that surged in the hearts of the women who stood firm by the trees they were protecting. In considering the Chipko story, we see that we must not mistake rage for violence or outward anger; rather we must see it as an intensification of energy that, transformed in and by community, powers change. These communities are not monoliths, but assemblages of people from different walks of life, socio-economic backgrounds, genders and faiths. It is easy to read stories of direct action and land defence and imagine a homogenous collection of people driven to anger, expressing it as one entity – a mob. Throughout history, our perception of what a protest is, has been influenced by the images we see in the news and in textbooks: thousands of heads weaving through the streets, marching or standing in unison. What is happening under the surface though is something more beautiful. There's a popular West African proverb that says it takes a village to raise a child; so too does it take a village, local and global, to transform the world through environmental action. Protests, marches, demonstrations and rallies, however big or small, are built by a constellation of unique individuals with varying skills, strengths and roles, without whom acts of resistance would be impossible.

In her essay 'Rebel with a Cause: How to Become an Activist' in the anthology *Nature Is a Human Right*, Noga Levy-Rapoport presents us with a better understanding of the roles and requirements necessary to create successful protests, rallies and marches. She splits these roles into two types: 'on the ground' and 'behind the scenes'. I find her framework incredibly important, not just in terms of amplifying and highlighting the sheer amount of work it takes to organise a protest but also in making them more accessible. For many, including those with disabilities or those fearing social and racial stigmas that make protesting unsafe, getting out onto the streets is not an option. There are thousands of people who are called to act and root in rage, but their action need not be confined to their physical presence at a march. As detailed in Levy-Rapoport's framework, protests need those who are skilled in administrative tasks – they are the glue that holds the protest together, who organise meetings, take minutes and onboard new volunteers. More recently, the success of a protest has become heavily reliant on its social reach, making the digital team's work incredibly important. People are needed to manage social media accounts, to communicate important information in online spaces and create hype around the action. And the importance of these types of role is not restricted to only the behind-the-scenes team. As important as social media reach is, it is also essential for protests to catch the attention of traditional media such as news programmes on TV and radio or newspapers, making spokespeople an essential part of a protest's organisation; these are people who are confident enough to speak to their town, region or nation about the goals and demands of the protest. An important role often overlooked when it comes to protest is that of the welfare and support team – the medics, the therapists and the cleaners. Those who make it safe for the protest to go forward and who provide support when there is trauma or burnout, which is all too common in direct action. The list could go on indefinitely and many organisations and platforms, including my own, have shared resources and frameworks to help make the myriad climate roles accessible to those who are struggling to find their place. Inspired by late US social change activist and principal organiser Bill Moyer, I created a resource entitled 'What's my role in the climate movement?', a collection of themes and related roles that can exist in the environmental movement. This work drew inspiration from Moyer's 'Four Roles of Activism' system, which he introduced in the 1970s to demonstrate how our roles in social movements are fluid and pluralistic. He wanted 'activists and movement organisations to understand that social movements require four roles and that participants and their organisations can choose which ones to play depending on their own make-up and the needs of the movement'.[16] He defined these four roles as 'The Citizen', 'The Rebel', 'The Change Agent' and 'The Reformer'. Moyers himself assumed many different roles, using his voice through journalism, broadcast television and activism. Although it was an instructive starting point, I felt that there was something missing in Moyer's roles and wider attempts at 'inclusion' from within the environmental movement itself. The roles lacked a sense of grounding in real-world stories, examples and scenarios. I extended Moyer's framework to non-exhaustive roles as follows: 'The Insider' (for example, lawyers, academics, politicians and economists), 'The Facilitator' (for example, key workers, nurses, cleaners), 'The Healer' (for example, religious leaders, doulas, spiritual guides), 'The Creator' (for example, artists, musicians, designers), 'The Visionary' (for example, technologists, sci-fi authors, philosophers), 'The Protester' (for example, land rights activists, land defenders and direct action activists), 'The Organiser' (for example, NGOs, social media communicators, facilitators, community organisers) and 'The Storyteller' (for example, writers, poets, elders, teachers). My framework was a deepening and extension of not only the different roles we can hold in environmental action but also the different ways we can connect to it. After creating this resource, I ventured further, thinking about the work of 'The Carer', mothers and growers who tend to and nurture this Earth and its inhabitants, or 'The Pollinator', those who bring together connections from different spaces and locations, building capacity and resilience and breaking down silos in the environmental movement. The more I explored these roles in my mind, the more I saw that regardless of using the label 'environmentalist', so much of the world was already engaged in environmentalism or had the potential to be agents of immense positive change.

Since Moyer developed his four roles, many valuable frameworks have been presented, including the 'Climate Action Venn Diagram' introduced by American Marine biologist Dr Ayana Elizabeth Johnson. At the centre of the diagram lies our role, sitting at the intersection of what brings you joy, what work needs doing and what you are good at. Another great framework presented by Deepa Iyer is the 'Social Change Ecosystem Map' that centres on equity, liberation, justice and solidarity, and identifies key roles, such as guides, caregivers or visionaries, that are needed to achieve those virtues at the centre.[17] More recently, the team at The Slow Factory, an award-winning platform creating solutions for social justice and human rights, released its 'Callings & Roles for Collective Liberation' framework which included generative roles such as the 'Inventor', who 'invents a particular process, system, culture or device that are good for people and the planet', or the 'Artist', someone who 'inspires people to be in touch with their humanity'.[18] Related specifically to the roles needed within activist groups was climate activist Tolmeia Gregory's bold and engaging series of posters entitled 'There is a role for everyone in the climate movement'. It lists the intimate and often unseen ways people are needed within, and are essential to, the movement, such as the tea-maker, the transcriber, the flag-maker or the photographer.[19] Each framework, resource or tool approaches action in the face of social and environmental challenges in its own unique ways, but what is clear is that in the wider climate movement, a kaleidoscope of roles is needed for impactful action. A myriad of openings for us to channel our rage in ways that are valuable and align with our strengths, skills and passions. What each of these frameworks is trying to communicate is the need for a sense of fulfilment and enjoyment in the environmental movement. They acknowledge that we each offer a unique set of skills fuelled by our individual passions. By acting in line with those passions, we can *joyfully* partake in action. For me, and many environmentalists I know, this has been the key to sustaining our action. Being able to show up in the ways that centre pleasure allows us to better contribute to the movement and reduces the risk of us burning out. As adrienne maree brown writes in her incredible book *Pleasure Activism*, 'pleasure is the point', and by rooting our action in it, we do the work to 'reclaim our whole, happy, and satisfiable selves from the impacts, delusions, and limitations of oppression and/or supremacy'. In the next chapter, we will explore how the absence of pleasure and the domination of doom in our movements accelerate and underpin destruction, and we will learn why joy is the inseparable and irreplaceable sister of rage.

6 Joy Is the Sister of Rage

In March 2023, I had the pleasure of sitting on the pre-show panel of the stage adaptation of International Man Booker Prize-winning novelist Olga Tokarczuk's *Drive Your Plow Over the Bones of the Dead* at the Barbican, exploring 'the Ethics of Activism'. One of the many nuggets of wisdom I gained in that panel discussion came from Jojo Mehta, the co-founder of Stop Ecocide International. When talking about anger and the despair that drives so many of us to care about environmental issues, she made the crucial point that, though the emotion is valid, we cannot stay in anger. We cannot reside in it. That anger is necessary but must at some point be transformed into action. Action sustained by anger alone is quick to burn out. The flame burns bright but not for long. What then can we use to sustain our anger and transform it into action without losing the spark? Joy. When we use joy as a tool to achieve transformation, we prepare ourselves for lifelong sustained engagement with the movements and campaigns we are so passionate about.

All too often, especially on social media, rage is not turned into action but doom. People are rightly angry and frustrated, scared of the world that is to come. This rage, as we have seen, is not only valid but necessary. Each year seems to bring with it a more intense deluge of news stories documenting the effects of the climate crisis on communities all over the world. From severe droughts to destructive fires and overwhelming floods. In these times, it has become ever more difficult to find hope. To feel positive about the future of the planet. To feel that we can carve out spaces of joy when there is so much destruction unfolding around us. Yet, only focusing on negative news does a huge disservice to all those who have worked and continue to work extremely hard to be part of the solution. All those who face adversity and still choose hope (through overwhelm, burnout, fear and helplessness) to create, build and dream better worlds. Like the sixteen young American plaintiffs from Montana who, in the summer of 2023, won, against all odds, a landmark court case with the judge ruling that the state had violated their constitutional rights by promoting fossil fuels and restricting their right to a clean and healthy environment. Or the 5.4 million Ecuadorians who, in the same week as the Americans' monumental victory, ushered in an equally huge and historic event by voting in favour of halting the drilling of oil in the Amazon rainforest. This was one of the most decisive democratic victories against the fossil fuel industry in Latin America and, arguably, the world.

Rage practised without a view of liberation, without a vision or embedded ritual of joy, becomes a dangerous weapon. Trading in doom for joy is not frivolous or a ploy to make people feel better about their inaction. Holding on to and cultivating joy through our rage is essential because so often doom comes at the expense of the most marginalised communities. As rage can transform from a motivating source of power into an engulfing, destructive force, so too can doom quickly, and worryingly, drive violence and harm. Some of the most cited pieces of alarming climate news appear in far-right publications and discourses, twisting statistics to blame immigrants, people of colour and the poor for the climate crisis. Over the last few years, we have seen these sentiments become the cause for tragic mass murders in Buffalo, New York, El Paso, Texas, and Christchurch, New Zealand. We also see these ideas perpetuated through predatory population control projects and organisations that, under the guise of women's empowerment, focus on controlling the population growth of Black, Brown and Indigenous women around the world. This is ecofascism – 'a fascist politic or a fascist worldview that is invoking environmental concern or environmental rhetoric to justify the hateful and extreme elements of their ideology', – much which originates in the Global North, which is responsible for 92 per cent of excess global emissions.[1, 2]

Just this knowledge of the cruelty of those who fatally express their anger and grief against others is enough to send many into the safe confines of detachment and apathy. But for me, it is this deep, terrible darkness that makes brighter and more inspiring the long history and continued efforts of the environmental movement and the resilience of the planet itself. I remember feeling a sense of palpable relief while reading on joy in the book *Braiding Sweetgrass* by Robin Wall Kimmerer. Writing on joy she reflected, 'Even a wounded world is feeding us. Even a wounded world holds us, giving us moments of wonder and joy. I choose joy over despair. Not because I have my head in the sand, but because joy is what the Earth gives me daily and I must return the gift'. That we need to think of joy as not only something we take from the world but also as something we are tasked to bring to it is an important realisation. Centring joy in our action is not frivolous despite the assertions of die-hard climate activists who recoil at the concept, convinced that you must give all parts of your soul as sacrifice for environmental campaigns. With Kimmerer's words we see that an absence of joy is an act of violence on the Earth itself; it is to ignore the beauty that the world offers us every day, without fail, and despite destruction. Together, this planet and joy sustain us, and in the words of the writer, poet and wildlife biologist J. Drew Lanham, we must remember that 'joy is the justice we give ourselves'.[3]

7 From Rage to Imagination

In this section, RAGE, we have explored how we might root in rage as a way to build and honour a natural connection with the living world and with our communities. We have learned that anger when confronted with destruction and oppression is necessary, but only becomes generative when it is transformed into action. Through the stories of marginalised communities – the Warren Country residents, the Ogoni 9, the quilombo, and those who led the Chipko Movement – we have learned that rage can take many forms. We have learned that we must use it to ignite and sustain us. That we should use it to fuel action both loud and quiet. That we must not only enact rage but embody it too. We have learned that this rage cannot be left to rot inside of us, making us hard and apathetic, but that to move us from despair to liberation, it must be fused with joy. The joy of connecting to sacred and beautiful lands; the joy of being able to breathe fresh, unpolluted air; the joy of being able to live without fear and conflict. As the writer, activist and environmentalist Rebecca Solnit makes clear in her book *Hope in the Dark*, 'Joy doesn't betray but sustains activism. And when we face a politics that aspires to make us fearful, alienated, and isolated, joy is an act of insurrection.' Accessing and enacting joyful rage is a rebellion, against not only the material impacts of environmental breakdown, but also the mindsets and worldviews that give rise to it. Seeing joyful rage as a way to connect to the living world requires us to be able to not only identify and fight for the systems we want to end, but also imagine those that we want to create. In the words of adrienne maree brown in *Pleasure Activism*, 'All organizing is science fiction . . . we are shaping the future we long for and have not yet experienced.' In the next section, IMAGINATION, we will reckon with the ways in which our ability to dream up new worlds has been limited by capitalism and colonialism, explore how these limitations tangibly impact essential ecosystems and learn from the wisdom and practices of Indigenous, and previously colonised and enslaved communities on how we can imagine the worlds our hearts yearn for.

Imagination

8 Reclaiming Our Collective Imagination

The year is 1856. It is not yet dawn and the forest is silent. It is a time of transition. The deep wash of night slowly fades, leaving you with only the anticipation of the morning. You sway back and forth, stomach churning, adrenaline rushing. A hearty greeting from the wind. At 80 metres above the ground, perched on a narrow but surprisingly strong tree branch, you have reached a dizzying height. Below you stretches an infinite vast landscape of complexity and wonder. Following, like a map, the deep-set grooves and ridges etched into the supple chestnut-brown bark of the tree over millennia, you descend. Down into the safe embrace of the forest canopy. A flash of grey startles you. Swooping, diving and curving it finally perches on a nearby branch. It is the Clark's nutcracker, a small, shrewd bird that uses its dagger-sharp beak to prise open large seeds from the surrounding pinecones, settling each one expertly beneath its tongue, to be then buried for the winter.[1] *You descend further, passing by the rare mountain yellow-legged frog, to the forest floor. You have reached the base of this mammoth tree: the giant sequoia. The Mother Tree. The ground is dry and soft, a bed of pine needles and small branches, prime fuel for the fires that catalyse regeneration when the ecosystem is kept in balance and pose one of the biggest threats to trees like the sequoia when the ecosystem is degraded. Around your feet is an incessant buzzing; small gnats (Azana malinamoena and Azana frizzelli) dart back and forth indulging in a breakfast of mushrooms, to be later consumed themselves by salamanders, spiders and bats. Minuscule and only recently understood, the gnat plays an essential role in the balance of this incredible ecosystem.*[2] *You crouch, eager to observe the clusters of mushrooms, dozens of waxy caps and Zeller's boletes, that adorn the lowest visible levels of this giant tree. If only you could see below the soil. Peer into the equally vast world of the mycorrhizal fungus network that belies the small size of these important mushrooms. You sit and ponder this dream, leaning on the Mother Tree, solid and unmoving. You do not yet know, and could never imagine, that in less than two centuries scientists will develop transparent soil so that they can study the systems you so desperately want to observe. You are also oblivious to the plots and schemes of the opportunistic settlers on this sacred land, those hungry for profits, who will succeed in the ultimate downfall, the death, of this beautiful tree.*

Hours pass, and as you rest your head on a bed of pinecones, another thought comes to mind. What if you could not only look beneath the soil, tracing the mycelium connecting the roots that spread hundreds of feet from the trunk to the other giants within the grove, but also see through the thick layer of bark to the centre of the tree itself? Of course, you could, as many have, cut cross-sections of the tree and admire the spongy phloem that transports sugars and nutrients and the thin cambium that reproduces to add to the tree's girth. This view would allow you to see past the growth rings of the tree, marking hundreds and even thousands of years of life, into the heartwood, the tree's centre. No. This is not the view you are after. You want to see inside *the tree. You want to know what it would feel like to stick your head through each of the Mother Tree's layers and observe its inner workings. What if you could visualise the conversations between arboreal friends and family through their interconnected roots or witness the flow of individual water droplets, ascending from the roots through the trunk to the leaves above?*

Your eyes gaze skyward and are met by a seemingly 360°

kaleidoscopic view of strong warm-toned trunks pointing inwards to a centre of criss-crossing branches and leaves. This viridescent crown filters warm shafts of light that bathe your face. Bliss. What bliss. You close your eyes, savouring the sensation. For a moment, it's as if you can still see the tree's branches behind your closed lids – an illusion cast by your optical blood vessels. This phenomenon shifts your train of thought: 'If, through illusion and other means, I can see a tree when it is not there, how else might I be able to use my mind to make connections with the rest of the natural world?'

'How much of this forest is left unseen when my eyes are wide open and how much more might I discover when they are tightly shut?'

In the summer of 2022, on a balmy August afternoon, I travelled from Cambridge to London to try and catch the last of the *Our Time on Earth* exhibition at the Barbican, where an installation of videos of myself and other incredible environmentalists were being shown. I was excited to head into the exhibition space and explore the creativity and imaginings being displayed. Walking into the sprawling space, I was met with a floor to ceiling visualisation entitled 'Sanctuary of the Unseen Forest'. With an image of buttress roots curving skywards, the visualisation was a compilation of thousands of intricate lines and dots in white, blue, red and yellow representing the vast number of interactions between water, soil, air and tree. The image slowly transported us from within the tree to outside of it, and back in again. Now a coating of vibrant green moss, lichen and leaves, a darkening expanse of forest canopy dominating the background. Then, a shedding of this outer layer, revealing the ceaseless activities of respiration, transportation and evaporation. This was my first encounter with the work of the 'experiential art collective' Marshmallow Laser Feast (MLF) and I was entranced. Transported into the body of the Kapok tree (*Ceiba pentandra*), a neotropical species found in Central and South America, Mexico, the Caribbean and parts of West Africa.[3] As described by the collective themselves, the visualisation was an experiment in 'expanding our sense of self beyond our bodies, reaching out into a sensuous experience' of the tree.[4] It was an exploration not only of the many minuscule interactions occurring within the tree, out of sight of our oblivious eyes, but also of interactions between ourselves and trees, which can be thought of as 'an extension of our lungs'.[5] The work of MLF allows us to reimagine our connection to the natural world and to uncover that which we miss, neglect or ignore in our busy lives. Through a series of virtual reality and visual experiences based on in-depth and hands-on research in forests in the UK to Colombia and by harnessing the power of LIDAR (a form of laser image processing), drones, CT scanning and 360° cameras, MLF has ushered in a new era of art- and technology-led conservation.[6]

It was the collective's 2016 project 'Treehugger: Wawona' that inspired the short fictional piece on 'nature's cathedral', the giant sequoia tree, which introduced this chapter. In this experiential installation, you are invited to hug, literally, a model of the tree, embedding your head within its trunk, and with the help of a VR headset, follow the journey of a single water droplet from tree root to tip. I, unfortunately, did not get the chance to visit this installation in person, but watching the video trailers uploaded by MLF, it's hard for my mind to not invoke its own projections of this experience. A slideshow of yet unexperienced memories. Of limbs, forehead and chest placed on tree. Of the messages and teachings transmitted from bark to skin. Through their unique work of art and visual storytelling, MLF connects people to environments across the globe in a way that is not sterile, detached or voyeuristic, but that evokes a sense of deep wonder and awe; feelings that have lain dormant in many of us since childhood.

As children, writing or visualising made-up scenarios to better understand the world, as I did at the beginning of this chapter, would have been a piece of cake, our minds playgrounds of imagination. As a child, I felt overwhelming curiosity about the world around me, which continued in my teenage years. That which I could not explain, I imagined, conjuring up the answers in my head. You might remember, too, how you held a multitude of ideas in your mind, turning a branch into a wand, grass into lava, mud into pies and clouds into animals. As adults, it is only too easy to dismiss these imaginations as trivial.

> *We cannot see into trees, what an absurd notion*
> *and waste of important brain power, are*
> *there not emails to attend to?*

As a collective, we are losing our ability to imagine – to imagine different ways of living, different systems to live by and different worlds to exist within. With the gradual democratisation of the means of distributing information on environmental breakdown, we now know more than ever before about the origins, impacts and trajectory of the environmental crisis. This information holds a potential for spurring on action. A potential for triggering reinvention. When we know more, we can do more. Awareness is an antidote to apathy and knowledge is power. Right? Then why is it that we – mainly those of us who live in the Global North, most notably those who hold positions of power and 'leadership', armed with statistics and facts – have made little meaningful progress on climate and environmental action? Why do those in the environmental movement continue to engage in the perpetual dance that is the parading of their own intellectual, political and moral standings? Why do they belittle those who are not embedded in or energised by their specific perspectives? Why are they motivated by the assuredness (and arrogance) that their perspective alone is 'right', that their circle of knowledge is of the greatest import?

> 'Carbon capture will save us.'
> 'Of course, it won't; how incredibly out of touch can you be?'
> 'We just need to return to how we lived in the medieval period.'
> 'Well, if we just curbed population growth, all our problems will be solved – there just aren't enough resources.'
> 'That is the most eco-fascist viewpoint to hold. Don't you know that it is about overconsumption of the richest on this planet, not overpopulation?!'
> 'The single most impactful act you can undertake to save the world is veganism.'
> 'Hey, why are none of these idiots talking about nuclear?'

And on and on it goes. Aided by polarising algorithms and an incapacity for nuanced dialogue, we have entered what the award-winning writer Amitav Ghosh calls the Great Derangement, an era that future generations will look upon with stupefaction.[7] An era in which the human species 'congratulates its self-awareness' but fails to carve out the time or space to dream and then most importantly *act*. We are all, in one way or another, guilty of fuelling this Great Derangement, or of observing it, quietly shielded by our phone screens. Our minds are full, our hearts are heavy, the problems grow larger, and our worlds become smaller. In the chaos of trying to make sense of a world that is crumbling around us, our capacity for imagination slowly,

quietly,

   imperceptibly,

          diminishes.

But is time not running out? Is it not reckless, dangerous even, to waste precious time on the follies of the mind? Must we not engage in pragmatism and logic, find the most scalable solutions and put them in place as soon as possible? Is the world not too chaotic and ever-changing to not be reactive? Mustn't we instead take to the streets and put out every fire that threatens to scorch us all?

Without the pause that imagination demands of us, we fail to acknowledge what Indigenous researcher Yuria Celidwen, a descendant of the Nahua and Maya peoples, knows to be true, that 'places of chaos are pregnant with possibility'. Investing time and energy in imagination is not frivolous, it is not just fancy, a notion Merlin Sheldrake, biologist and author of *Entangled Life: How Fungi Make Our Worlds, Change Our Minds, and Shape Our Futures*, strongly rejects. Speaking with him over Zoom, he tells me that 'imagination is one of the most important things that we have as sensing, feeling, purposeful beings . . . fundamental to how we apprehend the world'. What we think, materialises, and we bring the worlds we imagine into being every single day. The state of our current world and the potential for more destruction are direct outcomes of the systems we have collectively imagined since the dawn of the industrial era. Imaginings of wealth, of growth, of power and of consumption have manifested in the form of the climate and environmental crises. Our task now is to collectively envision a new way forward, not only as an exercise in practising fiction but as a radical act of rebuilding and remaking our world. We must take hold of and root in our imagination, which can guide us away from extractive, power-hungry systems and provide a blueprint for social, economic and ecological harmony. Without it we cede power to those – Western governments and leaders, billionaires, multinational corporations and the fossil fuel industry – who have clear imaginaries of the world they value, worlds that will continue to be built on greed, exploitation and destruction.

The climate emergency is not just a crisis of facts, figures and statistics, it is also a crisis of emotion and narrative. We are narrative beings. From the moment we leave the womb to the time we take our last breath, we are enamoured by stories. Whether through books, plays, music or art, storytelling helps us understand our place in the world and connect more deeply to our collective existence as a species. There are stories, new and old, to be told. Stories that can reach the hearts and minds of those who are ready for a change. When we look towards a more just future, we see that some of the most important and moving work will be in the arts and creative spaces: in the music halls, galleries and pages of books. In 2022, human rights and climate justice activist (and ex-international executive director of Greenpeace) Kumi Naidoo authored the 'Open Letter to the Philanthropic Community: Harnessing the Power of Arts and Culture for Humanity's Survival'.[8] The letter was a call to arms that spoke to a community of people – overwhelmingly those with high net worth and sizeable family trusts, who hold a huge amount of financial and political power – to pool their resources and invest not only in the techno-centric solutions being positioned by governments, but also in people. Naidoo goes on to tell us that 'cultural innovation in its deepest sense can help us answer questions about how we can live together, how we could radically rethink, renew and reimagine a better, more just world'. When we meet people where they are through practices of arts, culture and imagination, we harness the 'potential and responsibility for reimagining a better future'. I find this quote incredibly powerful, especially as a young person who witnesses all too often the patronisation and condescension of wealthy (often white and often male) journalists, politicians or commentators who proclaim so assuredly that we are doomed. Who scoff at expressions of hope or stubborn optimism. Who declare technological progress and continued extraction and consumption (only greener), next to simply giving up, as the only solutions available to us. As the Indigenous Guamanian writer Julian Aguon reflects in his book *No Country for Eight-Spot Butterflies*, their perspective represents the epitome of 'the language of limitation', a language we have become frighteningly fluent in. Might it be that these individuals benefit from the status quo, eager to hold on to the systems that have always protected them? Maybe they fear too much the unknown of a future that lies outside business-as-usual; they are not imaginative or industrious enough to believe that the unknown could be better for the majority of people on this planet, or they are too selfish and greedy to let that come to pass. Might they be the personification of the suppressive acts of colonialism which, as Aguon goes on to write, keep our dreams hidden away in tightly shut boxes?

We are living through a time of transition: The current world, rooted in oppression and exploitation, though not dead, is struggling to make way for the future world, rooted in justice and abundance. A time when the futures of our wildest dreams and worst nightmares have the potential to be brought into existence. It is with this perspective, through this lens, that we can acknowledge the urgent need to strengthen our capacity for imagination. This section is an exploration of the acts of those who dedicate their lives and passions to cultivating spaces for and practices of imagination – artists, writers and protectors of cultural and collective memories, working within the environmental space to reignite a sense of awe and wonder about the natural world, its unending complexity and enchanting beauty, and to bolster our journeys of action. As we go forward, we will hear from a constellation of thinkers, artists and writers who are committed to communicating compelling narratives, facilitating the future through their minds. From Fehinti Balogun, the playwright, actor and creator of *Can I Live?*, a digital performance about the climate catastrophe that premiered at the UN Climate Change Conference in Glasgow (COP26), we learn that dreaming is a form of planning. Along with the award-winning Icelandic-Danish artist Olafur Eliasson we will witness change and ask whether trees need therapy and if glaciers mourn their deaths. Delving into the legendary work and worlds of climate fiction authors such as Octavia

Butler, we will find that conjuring visions of the future through literature provide us with both warning and inspiration. Through the words of such authors, we will reimagine environmental action as an immense opportunity, as the driver of renewal, and acknowledge that embedded within the chaos of environmental breakdown lie not only endings but, crucially, beginnings.

In order to thoroughly examine our relationship, or lack thereof, to imagination, and how this hinders our natural connection, we first need to talk about the big Cs – colonialism and capitalism. In the next chapter, I trace the limits the big Cs have put on our minds and begin to ask what a decolonised environmental imagination may look like.

9 Decolonising Imagination

Colonialism and capitalism are holding our imagination captive. Our loss of imagination – though it may feel like a subconscious act – has in many ways been intentional, resulting from the influence of colonisation, industrialisation and capitalism. Those who uphold and profit from these systems, have an active interest in making it harder for us to be politically and creatively engaged. As African American author Saidiya Hartman tells us, 'So much of the work of oppression is policing the imagination.'[1] Many of us are worn out by long working hours, meagre pay and rising costs. We are overwhelmed by – yet also sensitised to – violence, destruction and despair fed to us through a polarising, biased and intense news cycle. We are lured into a culture of consumption, and a deluge of products and technologies are marketed to us with the assurance that they will make us cooler, more likeable, more productive and more fulfilled. And when it is not being concealed, the destruction and exploitation involved in the manufacture of these products is convincingly presented to us as necessary and essential. Take as an example the boom in electric vehicle sales. Instead of advocating for and building policy to support affordable and efficient mass transportation systems and cut down car usage, the agents of capitalism are convincing consumers that shifting to electric vehicles is the solution to the fossil fuel problem. What the electric vehicle industry is really doing is furthering exploitative practices such as alleged land grabs in Indigenous territories like Peehee Mu'huh in Nevada – where the tribes say lithium mining has been approved without proper consultation with the Indigenous community who call the region home and whose ancestors were massacred there in 1865.[2] The electrification of our grid and transportation systems is an essential part of the green transition, and electric vehicles are a big part of moving towards the goal of net zero emissions. However, a green transition built on dirty materials and ethics is not in line with the philosophy and practice of environmental justice. As you will read in the section CARE, our transition to a greener world must be just. Allowing private manufacturers to profit at the expense of Indigenous, Black and Brown communities for the continued consumption and comfort of the Global North is not part of the vision for a just, safe, sustainable and regenerative future.

In the face of the destruction of Indigenous life, worldviews and lands, decolonisation becomes an essential tool of resistance. Decolonisation is a term most of us have encountered in reference to the lack of transparency about colonialism and imperialism in our school curriculums, with organisations like The Black Curriculum filling in the blanks of Britain's colonial past often neglected in most British schools. The meaning of decolonisation has evolved over the years, expanding from the act of a country, state or community becoming independent from a colonial power or reclaiming stolen land, to a wider act of challenging and dismantling the cultural, intellectual and social effects of colonialism that continue to be preserved and perpetuated today. In the wake of George Floyd's brutal murder by a white police officer in Minneapolis, Minnesota, in 2020 and the historic global mass protests that followed, decolonisation became somewhat of a buzzword. Despite the independence won by African and Asian countries between 1945 and 1960 and decades of campaigning by activists, educators and communities to unearth the violent and oppressive histories of Black, Brown and Indigenous communities, it seemed that in 2020, the world finally woke up to the pervasiveness of racism, oppression and coloniality in modern systems. Decolonising practices and programmes quickly spread, tackling Eurocentrism and systemic oppression in fields ranging from academia, politics and the law to conservation, the arts and sport. But while significant effort, and progress, has been made in many areas in this regard, when it comes to imagination, the claws of colonialism and capitalism grip society so tightly that we are convinced there is no alternative. No alternative to viewing nature and people as resources to exploit, to wring dry until we have extracted the last drop of profit. These claws leave deep wounds, filled with fear, anxiety and exhaustion, and the systems that create them escape without a scratch. We are unable even to imagine or allow others to conceive of any alternatives. To paraphrase the Anishinaabe and Ukrainian writer Patty Krawec, we are first colonised when our collective sense of a future is taken away.[3]

Racism, extraction, pollution, destruction, poverty and overconsumption are tools of imagination wielded by colonialism and capitalism to maintain the status quo. They keep us from seeing the world as it was and as it could be. Their effects can be seen in the way in which many Indigenous languages are now on the brink of extinction as they were banned from being learned or shared in the colonial era and in the loss of knowledge that has led to. Or in the way Indigenous approaches to conservation were suppressed for centuries, making way for colonisers to establish dominance over the natural world. Colonialism and capitalism push the possibility of a symbiotic worldview, one that aims for harmony with and mutual respect for nature, out of our imagination.

But whilst capitalism and colonialism have reigned

supreme for much of our recent history, in comparison to the knowledge systems of Indigenous communities around the world, they are infantile. Through her short story 'The Long Memory', featured in *Octavia's Brood: Science Fiction Stories from Social Justice Movements*, social worker and writer Morrigan Phillips explores the power of collective memory, and hence imagination, when cultivated outside of colonial and Western frameworks of the world, and the real threat it poses to systems of oppression. The story opens with an impassioned speech from Councilman Holt, a powerful merchant-cum-government official undertaking an intense national campaign to warn the citizens of The Archipelago about 'The Long Memory': 'the most dangerous idea threatening our peace, prosperity and security'. It is an act or phenomenon so damaging that it holds back civilisation itself. We then meet Cy, a keeper of 'The Long Memory' and orphaned refugee. Cy is on a mission to unmask the true fear of Holt and his followers, the colonists and supremacists who will go to extreme lengths, often violent ones, to suppress 'The Long Memory'. 'You fear the people remembering', Cy exclaims, 'that they will remember you . . . killed many and burned the libraries, and they will ask why. And you will have to answer.' You see, Cy knows the power of collective memory. She is awake to the fact that 'a people who remember will not be exploited again. A people who remember will take action'. We must be those people, the people who remember and who are keenly aware that the way we remember and honour history – our collective memory – shape the way we imagine the future.[4]

I began Chapter 8 with a little thought experiment, imagining the exciting, daunting, intimate and what would be a totally hair-raising journey, scrambling down the trunk of The Mother Tree, the giant sequoia. In the nineteenth century, the story of the Me-wuk – indigenous to the land now recognised as Northern California – and the surrounding tribes, such as the Pomo, Sinkyone and Cahto, with whom they shared and continue to share the land and the giant sequoia trees was one of destruction and disconnection. The people faced brutal genocide. The land faced ecocide. The killing of both people and the land.

It was the era of Manifest Destiny, when (white) man was believed to have been endowed with the right and duty, by God, to dominate over and exploit the natural world. In a three-year period, from 1853 to 1856, two of the world's largest trees, the Mammoth and the Mother trees, would be lost to this doctrine. Cut down and toured over weeks in the US and UK. These practices of domination laid the foundations for acts we now define as ecocide, a term that became enshrined in law across Europe because of the incredible campaigning of the team from Stop Ecocide. It refers to unlawful or wanton acts that are committed with knowledge that there is a substantial likelihood of severe and either widespread or long-term damage to the environment that will be caused by those acts.

Both the Mammoth and the Mother trees were doomed, in an act of irony, to be lost to fires in New York and London. Ironic because sequoias have an intimate relationship with fire – for them fire equals life. The cones of a sequoia open and release the seeds within them only when conditions are dry enough, as after a fire. Not only does the fire eradicate close competitors, but also exposes the soil, and the important minerals held within it – making it the perfect home for brand new seeds. These facts were well known to the Indigenous communities who had lived with the trees for hundreds of years and conducted small, low-level burns to regenerate the sequoia ecosystem. A decade after the destructive fellings, John Conness, the Californian senator at the time, vowed to pass a bill to protect what we now know as the Yosemite Valley. The region would not become a state park until the early 1930s but would influence the creation of other protected areas in the state, including the Sierra National Forest and Sequoia National Park. But with this protection came disconnection. A disconnection of communities with their kin. For the communities native to the land, the trees were and are not just trees but relatives, and the erosion of their cultural and ecological knowledge through barriers and the imposition of the law was devastating.

In recent decades, the Southern Sierra Me-wuk Nation has experienced the worst of the increasingly devastating forest fires on the West Coast of the US. Events that are now being referred to as *super fires*, which burn and obliterate everything in their path.[5] Yet the Me-wuk Nation are, at their core, fire people. Over centuries they – and many other Indigenous communities across North America as well as Aboriginal communities in Australia – have engaged in the practice known as cultural burning. It is a practice that allows Indigenous communities to live in a symbiotic relationship with their local environment and is an important part of sustainable land management as well as cultural connection. Cultural burning involves the intentional small-scale and controlled burning of forested areas to clear the forest floor of dried twigs and debris that would otherwise act as the perfect fuel for future forest fires. These small-scale fires are also regenerative; they build biodiversity and clear space for sun-seeking plants to grow, allow the foraging of the culturally important elderberries, and support Indigenous food sovereignty through hunting. For the communities that practise cultural burning, fire, in the right doses and with the right cultural embeddedness, is medicine.[6] But, with the colonisation of the region and the subsequent formal creation of many national parks, the practice of cultural burning was wiped out. The suppression of the region began in 1850 upon the instruction of California legislature; it marked the beginning of the decades-long murder of thousands of Indigenous peoples in California – what has been referred to as 'the clearest case of genocide in the history of the American frontier'.[7] This is the pattern of colonisation, to eradicate Indigenous people and their knowledge, and to suppress and dominate the Earth and its people. In order to enact this domination, marginalised communities must be threatened with violence and Indigenous ways of knowing must be devalued. There is a dissonance that clings, like a heavy stench, to this pattern of colonial imagination, one that is rigorously explored by the anarchist and anthropologist David Graeber and archaeologist David Wengrow in their ground-breaking book *The Dawn of Everything*. Their book documents in detail the main contradictions, lies and perversions of history charted by Western imperialist intellectuals in order to force capitalism – and their power – on Indigenous communities through slavery and colonisation. Through their analysis of a wide range of case studies, the authors ask why the dissemination of Western civilisation, if it was so superior, required so much violence and manipulation. Importantly, their book urges us to 'ask better questions'. To challenge Eurocentric and West-dominant perspectives and interrogate whose version of 'reality' we are living in and upholding. To ask these questions requires us to see beyond the limits of what is 'known' and what we are told. Only by untethering our minds from the limits of an underdeveloped, arrogant and penetrating colonial imagination can we escape the grip of social and ecological catastrophe.

Over the last decade, the West, while continuing to perpetuate colonial and capitalistic approaches to environmentalism, has become intrigued by the worldviews and practices of Indigenous communities in relation to environmentalism. In recent times, cultural burning has become a formalised land management practice in Australia and North America in what is seen as a blending of 'Indigenous knowledge and Western science' – a distinction and separation that encodes ideas about who in this world practises 'science', forgetting that Indigenous knowledge *is* science. Although it is great that Indigenous methods are slowly gaining respect in the environmental space, it is heartbreaking for communities to have been silenced for so long and to witness the reverence shown to Indigenous communities often turn to romanticisation and tokenism. Indigenous ideas and practices make for good stories to be marvelled at at conferences and dinners, but the deeper demands of reparation, returning of land, dismantling of systems of oppression and higher Indigenous representation in decision-making spheres remain too radical to be considered.

In Britain, a country ever eager to parade its pride in its history, lack of connection to ancient Indigenous British environmental practices is astounding. The land here practically trembles with ancient folklore, wisdom and sacred knowledge that lie buried in the soil with ancestors. In the late summer of 2021, I spent a twilit night looking up at the moon in the Kent countryside. Snuggled on a hay bale with a new-found friend and surrounded by a wonderful community of dreamers and weavers. We had gathered for a two-day retreat convened by *Emergence Magazine*. That evening, Mercury Prize–winning folk singer and author Sam Lee sang to us, beneath the moon, the songs he had been taught, and which he now preserves, by Scottish and Irish travellers, aggressively marginalised and isolated in the UK. Songs that tell the tales of woodlands and rivers, of harvests and moons and of the bird he treasures, the nightingale. A bird whose numbers in Britain have depleted by 90 per cent.[8] The songs Sam sings honour and bring to life the ancient teachings of this land – they are an act of remembering.

We must remember the Druids, who since the European Iron Age have practised an Earth-honouring, animistic way of life. In a conversation with Lucy F. Jones for an essay written for *Emergence Magazine*, the elderly Druid Damh the Bard describes the druids as a broad and disparate community who see the living world as a sacred book that acknowledges existence on this planet as a cycle of life, death and rebirth. The druids practise the nurturing of not only the outer but also the inner forest, honouring the presence of the living world within us. We must also remember the Celts, who regard women as guardians, keepers and leaders of the natural world in their mythology. Women like Cailleach, who formed the mountains of Scotland and Ireland and considered the crow and the deer as her sage advisors.[9] Britain has a history of deep kinship with the living world, yet we do not hear calls for re-establishing these ways of being at big climate conferences or by decision-makers.

From Brazil to Indonesia, Britain to Kenya, there are practices Indigenous or not, ancient and contemporary, that can re-enchant and re-locate us in this damaged landscape we find ourselves in. Practised ardently by few, they wait to be respected, to seed a natural connection between ourselves and the planet we call home. When Robin Wall Kimmerer, in her bestselling book *Braiding Sweetgrass*, called us to 'become indigenous to place', she asked us to care for the land 'as if our lives, both material and spiritual, depend on it'. That we are allured by Indigenous practices and traditions indicates our deep desperation. We might not have fully accepted it, or looked it square in the eyes, but we are ready for change.

The resurrection of or reconnection to our individual and collective imagination cannot be considered through the dominant Western lens. We do not have to inherit systems of oppression. The perpetrators and upholders of the colonial agenda of disempowerment – right-wing governments, multinational corporations, rigid academic institutions and fossil fuel companies – recoil from those who dream of better worlds, quickly dismissing their ideas as too fantastical. They slash down these dreams with the sword of *reality*, forgetting that reality is something that we make, something that is malleable, slippery and ever evolving. We forget that reality is a shapeshifter.

Many of us, though not all, are able to travel through space and time when we close our eyes at night, taking for granted the marvel of our minds, of the power of dreaming. But how many of us consciously bring this magic into our lives when we are awake? Maybe we are still traumatised by the scorn and embarrassment of being caught daydreaming, hypervigilant about averting the judgemental eyes of our peers or bosses. In the imagination of the West, dreaming is suppressed and branded as unproductive;

however, for many non-Western communities, dreams are seen as parallel realities. They decode and take seriously the messages from dreams and apply them to the material world. The ability to blur the lines between waking and dreaming allows us, as it did for those living under oppression, 'to liberate [our] imaginations from the ... narratives instilled by colonial educational regimes ... to connect to the universal struggle for liberation'.[10] This dreaming is not unproductive aimless thought or fantasy. It's not about moving through the world asleep or unconscious, detaching from the demands, ills or pains of our society. As Fehinti Balogun told me, 'Dreaming is a form of planning and when we give ourselves ceilings and shackles, we limit our dreams to the sphere of what is allotted to us.' Collective dreaming in this sense is about asking *what could be otherwise*, dreaming it, sharing that dream with others and coming together to build that vision, brick by brick. It's about paying attention, understanding and critiquing the systems that do not work for us and creating space for new ideas for living to emerge.

By decolonising our imagination, we resist the limits that have been imposed on the possibilities for better futures by systems that profit from the Earth's continued plunder. We must come to know that our imagination is a tool of resistance that, when sharpened collectively, allows us to extend the limits of the more just and abundant worlds we can bring into being. In harnessing and building collective dreams for the future, we are tasked with becoming radically awake to and in tune with ourselves, each other and the planet, embedding what we envision into how we show up in and change the world. One way we can do this is by learning from the communities who have known, lived and seen catastrophe before, people for whom the trauma of living through disaster has been woven, intergenerationally, through their DNA. Connecting to their stories of renewal, regeneration and reimagination gives us not only perspective but also no excuse. It forces us to witness and summon the power and courage that we need to remake ourselves and the world. But there also exists a wide network of teachers who cover nearly every square inch of the planet. Teachers who have known, seen and experienced change on scales both mammoth and minuscule. Teachers who have looked extinction square in the eyes, some recovering, others transmuting and still others returning their energy to the Earth to make new again. These teachers are the rivers, the rocks, the mountains, the trees, the lichen, the foxes, the birds and the fungi and so on. The non-human living beings we so often forget we share this planet with. In the next chapter, we will ask whether glaciers mourn their own deaths, opening up a conversation about animism and how we might bring this ancient, diverse and world-spanning practice and perspective into our contemporary lives.

10 Does a Glacier Mourn Its Death?

In early February 2023, I had the honour of virtually convening with the Icelandic-Danish artist Olafur Eliasson, who took me on a brilliant journey through topics across ecology, philosophy and art. Eliasson's career spans three decades and his work has had a global impact. From designing the 2007 Serpentine Gallery Pavilion to being appointed the United Nations Development Programme Goodwill Ambassador for Climate Action and the Sustainable Development Goals, he blends creativity with purpose and action. One part of our chat that stuck with me was a recounting of a previous conversation he had had with Natasha Myers, the associate professor of anthropology at York University, Toronto, who coined the term 'planthropscene', which explores how humans can heal their relationships with the natural world.[1] Eliasson describes how in his first conversation with Natasha, she told him that she had just come from tree therapy. To that, he replied honestly, that he had no idea what tree therapy was, but that he was curious to learn more. He asked whether this tree therapy had something to do with the energising feeling we as humans receive when we encounter trees. No, she said. It was not that. She was returning from a session where she was *giving* a tree therapy. Myers was responding to the trauma experienced by a tree that lived near her house. Having once existed within a grove, a community, of other trees, it was now alone with the others having been removed to make way for a new road. Eliasson told me this story in response to a question I had asked on his connection to and practice of imagination. He finished the story in an honest and reflective way, sharing how struck he was at his lack of imagination – 'I could not imagine that the tree was the main agent', he said. If I am honest, I think I would have reacted in the exact same way. The story made me reflect on how even those of us who feel deeply connected to the natural world spend so much of that time centring ourselves. We show disdain for extraction and destruction in natural landscapes yet take all we can get – spiritually, emotionally and physically – from these spaces without giving back. It was heartening to hear how Olafur had been inspired not just 'as an artist, but as a scientist and feminist', and had started to think about the role of art and culture as a way of imagining a world where humans were decentred.

As I sat reflecting after my conversation with Eliasson, I wondered whether, if unknowingly, his art already evoked a view of climate breakdown where the human was decentred. Flicking through the mammoth archive of mesmerising and impactful work on his website, I was caught by an arresting picture of Eliasson in 1999 standing on a stretch of tarmac, its black, tar edges blurring into the grasslands beyond. He stood beside a small white and red TF-1 carrier plane, having returned to his homeland to photograph a series of thirty glaciers as a part of his wider environmental artistic practice. He called this *The glacier series*. Two decades later, the artist returned to Iceland hoping to reconnect to and re-photograph the same glaciers. With the original photographs on his lap, eyes scanning the wide landscape and fingers poised to capture the perfect shot, Eliasson and his pilot were met with slight confusion. They had travelled to the exact same locations yet were confronted with a grossly altered view. They were witnessing the stark manifestation of the Anthropocene, the geological age we are currently in, where human activity is the dominant driver of ecological and climatic change. Many of the glaciers Eliasson had photographed earlier had melted significantly, some of them to the point of near extinction. And so was born *The glacier melt series 1999/2019* (2019), a sequence of photographs pairing the 1999 and 2019 photographs of each glacier. A memorialisation of the immense effect of our species on the rest of the natural world. The same year, as the second set of photographs were taken, a funeral was held for the Ok (Okjökull) glacier, once a mighty and slow being that stretched over vast swathes of land, now a giant ice puddle in the centre of a barren landscape. To commemorate the funeral, the writer and poet Andri Snær Magnason engraved a 'Letter to the Future' on a metal plaque placed at the base of the remaining glacier. The letter went as follows:

> *Ok is the first Icelandic glacier to lose its status*
> *as a glacier. In the next 200 years all our glaciers*
> *are expected to follow the same path.*
> *This monument is to acknowledge that we know what*
> *is happening and what needs to be done.*
> *Only you know if we did it.*
> *August 2019*
> *415 ppm CO_2*

What does it mean to hold a funeral for a glacier? What if we were to commemorate all that is lost each day? How many minutes of silence would we need to take, and would the overwhelming loss push us to action? For Eliasson, melting glaciers are evidence of our *inaction*. A mirror of our inability to use what we know to make positive change. So how is it that we translate what we know into what we do? For Eliasson, this is a question he centres in his work and one he sees art being able to answer. For him, 'Art is the distance between knowing and thinking.' A space with the potential to materialise the questions we do not yet know to ask, the paths we do not yet know to take, the connections we do not yet know to make. While many of us, including myself, will not have the opportunity to see first-hand the photographed glaciers in their 1999 glory, in the summer of 2019, I had the opportunity to get up close and personal with *The glacier melt series*, alongside many other provoking and beautiful works in Eliasson's *In real life* exhibition at Tate Modern in London. A reflection that *The glacier melt series* provoked in me was on the way in which such inherently slow beings as glaciers have been altered into becoming physical manifestations of the accelerative nature of modern human life. They have gone from colossal, seemingly eternal existences to complete extinction in only a fraction of their lifetimes. We overwhelmingly see this extinction as a threat to our own human-centred lives. We imagine freshwater shortages, rising sea levels, the 'end of the world'. Yet, what thought is saved for the glaciers themselves? (How) do they mourn their imposed disappearance? (How) do they cry out for the lives they once lived? (How) do they adjust to a bleak and disrupted present? While these questions may at first seem abstract, impractical or fanciful, to many cultures (past and present) around the world – from Aboriginals, Ecuadorians and even the Celts – who practise animism, they are instinctual. From the Yukon Valley to the Himalayas and the Peruvian Andes, glaciers are perceived not just as large hunks of ice, but also as those that 'hold a complex place in the human imagination... seen as malevolent forces... as sources of inspiration, as living beings, as the homes of gods, and even as the energetic ground from which all sustenance and prosperity springs'.[2] And as we are learning now, glaciers hold the story of diseases of the past, with scientists discovering 15,000-year-old viruses in ice samples taken from the Tibetan Plateau in China.[3]

I first engaged intentionally with the concept of animism during the first Covid lockdown in 2020 through an *Emergence Magazine* interview with the American ecologist and philosopher David Abram. Abram is best known for his work on the 'ecology of perception', the ways in which we sense and engage with the rest of the world. He strongly believes that we have lost the skill and ability to engage deeply with the rest of nature and to behave through our instincts in our interactions with the non-human world. For Abram, 'language and meaningful communication are not strictly human capacities' and he is certain that 'every living entity speaks, but that most humans have lost their abilities to understand them'.[4] Letting go of human-centred thinking is a mammoth challenge for many of us living in the Global North, who have been accustomed to the philosophical and religious teachings of the Western world. But what happens when we begin, not only to imagine, but to deeply witness and engage with the non-human world around us in meaningful ways? How does this shift embolden and root our environmentalism?

For many, animistic thinking remains just that. An abstract way of thinking that is hard to connect to. In my conversation with the writer Robert Macfarlane, we spoke about the difficulty many, especially those brought up in and with a post-Enlightenment worldview, have in 'reaching out of their minds' – as Ursula K. Le Guin would say – and rethinking the stories they have been told and the ones they tell. We met at a time when Robert was reflecting deeply on his journeys visiting rivers around the world for his upcoming book *Is the River Alive?* – which he tells me, 'It absolutely is!'. It was through these journeys that Robert came to better understand animism, and the aliveness of the living world. You may have noticed that throughout this book, I use this phrase 'the living world'. I first came across this phrase in Robin Wall Kimmerer's *Braiding Sweetgrass* and was inspired to use it as much as possible by Robert, who uses it as a replacement for 'the environment' or 'nature' as a way to acknowledge its aliveness in our language.

As we speak, he tells me that his process of truly witnessing the aliveness of the living world has been in the making for forty years. Only recently has it become clear to him, through his interactions with communities very deeply connected to animist worldviews, that the aliveness of the world is so obvious that it is boring. With this realisation he has been able to craft a model of how we might more keenly honour the spirit of the living world. Less of a model and more of journey, which would bring us to a knowing so engrained in our bones that, as Rob tells me, a 'river is alive, sentient, gives life, and enlivens in ways that exceed the sum of the lives that it contains'. The journey, he describes, would be similar to that of a secularist moving from outside of to within a cathedral. In the first stage of the journey, he begins, 'You might see a cathedral as a secularist and say, "I recognise that this building has been built to contain an idea", and then maybe you cross its threshold and enter into its atmosphere.' This, I learn, is stage two, where the generated atmosphere moves you, 'you can walk around and admire it, but you don't yet believe'. The next stage, unfettered belief, is one he found himself reaching after spending a long time with the Innu People and the Muteshekau-Shipu River that flows through Nitassinan, the name the Innu people use for their unceded territory in what is also known as Quebec. This stage was, for Rob, a time of deep reckoning, awe and reverence, reaching the unwavering, indisputable belief that the river was alive. The last stage brings us to where we began, where the aliveness of the Earth is so obvious it becomes an unmentioned yet central part of our lives – this stage is knowing, a knowing that is settled in our bones. For Robert, taking people on this same journey is one of the hardest but most necessary storytelling tasks he has set himself. It is clear that this is also one of the hardest tasks we must set upon ourselves. Journeys we must go on, whether along rivers in Quebec or our local streams and brooks. How do we traverse the deepening stages of reverence to the living world, becoming unshakeably sure of the aliveness of the beings we share this planet with? How do we extend this belief to the ecosystems and organisms close to home and those on the other side of the planet? Pushing beyond valuing only what we own – in our gardens and homes – towards a holistic knowing and connection to places that are prohibited, abandoned, polluted or seem invisible.

The question, 'is a river alive?', is a simple yet essential one. The answer of which shifts the way we treat the living world. In response to this question, the dominant Western perspective would have loudly and confidently proclaimed, NO. A worldview that would lead to the pollution and destruction of rivers around the world, including the Whanganui River in Northern Aotearoa (New Zealand), that in 1970, after 150 years of unbridled pollution, was close to being classified as dead. In Chapter 20, we will learn how as a result of 200 years of Indigenous activism this river was saved from extinction by reinstating the belief that the river was alive and that it deserved personhood, not only in the Indigenous worldview, but in the Western perspective and policy too.

Our work then, if our environmental action is to develop a natural connection to the living world, is to unsettle and question the imaginaries that drive our disconnection, and instead see ourselves in every river, rock and rainforest on the planet. To see the living world not as a passive backdrop but as a complex, dynamic and vastly intelligent being that has so much to tell us about where the world has been, and where it is going.

11 Parables for the Future

I want you to imagine an apocalyptic landscape, a land, say in North America, that has been decimated by climate breakdown. What do you see? A worn out, barren desert? Dilapidated fields scorched by an unforgiving sun? The last remaining humans, if any, spiritless and riddled with fever, on the doorstep of death? Humans who have suffered a scarcity of water, who live in constant fear of violence? What if I told you that this land is not one of our future, or a description of the set in the critically acclaimed show The Last of Us, *but of our relatively recent past? These descriptions are borrowed from the words of Abolitionist and writer Frederick Douglass in his 1855 book* My Bondage and My Freedom, *in which he described the post-slavery landscapes he witnessed. In its wake, slavery, and the intense plantation industry of the time, left Black people and the land depleted. For the communities across the Americas and Africa, the centuries-long transatlantic slave trade was an apocalypse of its own.*

Over the last two years, a small tradition has emerged for me and my husband, who I have been lucky enough to introduce to my homeland of Ghana. Squirrelling away a week or so outside of my research each year, we spend time with family and the various forests, valleys and beaches of the Ghanaian landscape. We walk, knees buckling, across the tree-top canopies of the Kakum National Park. We splash and squeal like children in the frothy, powerful shore dumps of Cape Coast's beaches. And every year, we tour, solemnly, the slave castles of Ghana's coastlines.

Slaves were trafficked from all corners of Ghana down to the coast to be stored in these castles, most famously Cape Coast Castle and Elmina Castle, which were owned and managed, over a series of changing hands, by the Portuguese, the Dutch and the British. Walking on the blood-blackened, cold and worn cobbles, ducking into dark oppressive dungeons and listening with rapt attention to the always erudite historians, the visits are always soul-crushing and enlightening in equal parts. I'll leave it up to you to spare some time to explore and engage with the history of these castles; for me, the connection between the castles and the forests remained an overwhelming discovery. Standing metres away from 'the door of no return', the gateway that all enslaved Ghanaians were led through to embark on the ships that would take them towards more suffering, violence and pain, never to see the forests of their land again. You see, many of the slaves taken from the Ashanti region of Ghana came from powerful states, including Adansi, home to the forest at the heart of my PhD research. Mine was a dizzyingly full-circle realisation of the inextricable link between the land, our connection to it and the destruction wrought by colonialism and capitalism, which uprooted my ancestors from their forests and sold them for the very gold that lay in abundant deposits below those very same trees. A realisation which prompted in me, over and over again, the same questions. *How, in the face of such catastrophe, were my ancestors able to continue, to go on? How could they have possibly imagined anything other than strife?*

*Could they,
did they,
imagine me?*

My existence, the existence of every Black person alive today and all those who have lived over the last 400 years, is evidence of the fact that our ancestors did continue. They did face catastrophe. They did endure strife. And with each generation, they imagined us. As Yale scholar A. J. Hudson writes:

> *The world has already ended several times. It ended the moment Columbus landed on the islands of the Caribbean . . . It ended for the kidnapped villagers in Western Africa—my ancestors—when they were stuffed into the wicked belly of a slave ship and cast into slavery in a strange land with a strange new climate. It ended with the Trail of Tears. Agent Orange. Hiroshima. The Holocaust. It ended with Hurricane Katrina, Maria, Kenneth, Harvey and Dorian.*[1]

And, as I write, it has ended for the 35,000 – and counting – Palestinians and their families, who have died in the ongoing genocide and it is ending for the 2.5 million Sudanese estimated to die from famine by the end of 2024.[2,3]

If in these darkest of times, communities of the past and present have been able to imagine liberation – amongst the harrowing images of death coming out of Gaza emerge videos of children, men and women dancing and singing, writing and sharing poetry – how then is it so difficult for those of us in the West, the majority of us living infinitely more safe and comfortable lives, to look beyond environmental breakdown and imagine liberation? When it comes to the climate and biodiversity crises, the stories we tell, whether real or fiction, are overwhelmingly dystopian. We've replaced *once upon a time* with *at the end of the world*, and our minds and hearts are suffering for it – as is the planet. Why are the stories we tell so bleak? Why do we find it so easy, so rewarding, to speak *the language of limitation*? Is our current collective task, that of ushering in a better future, not simply the task of rewriting our collective story? If we cannot imagine thriving futures in the stories we tell, stories that act as canvases on which we can sketch, test, invoke and inspire new worlds, then what hope is there?

For previously and currently oppressed communities, telling stories is an essential method of processing the horrors of their reality; but it is also a way to create sacred worlds in which they can be seen as they are, as human, as worthy. The stories we tell, both good and bad, have the power to transform, to come alive and manifest themselves within our actions in the physical world. The stories we tell can be prophecies, predictions and portals to future worlds. There exists a long tradition of storytellers who tell stories of the future from unimaginable presents. It is these writers that we can learn from, who balance radical authenticity and radical imagination in each sentence. Who, quoting Julian Aguon again, 'engage in the problems of the world without losing clarity about the futures we need to build', particularly those who contribute to the growing canon of Afrofuturist writing.[4]

Afrofuturism, as a term rather than a practice, was coined in 1994 by Mark Dery, a white American author, lecturer and critic, who in his essay 'Black to the Future: Interviews with Samuel A. Delany, Greg Tate and Tricia Rose' questioned the lack of sci-fi writing that featured or was rooted in the Black experience. You might have encountered

Afrofuturism through Marvel's mainstream hit *Black Panther*, a film and world we will return to explore in depth in Chapter 14. As defined by Dery, Afrofuturism is 'speculative fiction that treats African American themes and concerns' and asks whether 'a community whose past has been deliberately rubbed out can imagine possible futures'. As the Zambian writer and photographer Masiyaleti Mbewe observes, 'because of the physical, cultural and environmental disruptions of colonialism (the transatlantic slave trade, apartheid, neo-colonialism), Black people have always had to consider living beyond the end'.[5] And how beautiful, how utterly earth-shaking is it that despite facing the end and continuing to live amid 'apocalypse', Black communities continue to imagine past that seemingly finite point in space, crossing the end line over and over again. The past and present imaginings of the future, and of the predominantly Western future generations, have been born of a group of mostly white men encoding their own philosophies about the world and their connection to land. Take the popular 2014 science fiction film *Interstellar*, directed by Christopher Nolan: a film that as an aspiring astrophysicist and subsequent astrophysics undergraduate I found awe-inspiring. While I was entranced by the incredible cinematography, set design and soundtrack, certain elements of the film's script felt disconcerting. Cooper, a trained NASA pilot, is tasked to find a new home for the human species, who had obliterated Earth and were suffering from an ever-worsening food crisis. Sent into a fury triggered by the comments of his daughter's teacher, who was adamant that the children of Earth 'need to learn about this planet, not tales of leaving it', he proceeded to vent to his father. He lamented how we humans had 'forgotten who we are ... explorers, pioneers. Not caretakers.

We used to look up and wonder at our place in the stars. Now we just look down and worry about our place in the dirt'. We don't have to look far to see how narratives like this play out in the real world, with billionaires like Elon Musk emphatically embarking on 'the road to making humanity multiplanetary' and aiming to colonise Mars. These narratives deny our need, duty and innate capacity for caretaking. Narratives that cast the Earth as a hostile home rather than an embodiment of our own careless actions. Narratives that divorce us from a planet that we are so interconnected to that when we harm it, we harm ourselves. As the writer John Halstead writes in his essay 'The Most Dangerous Story Ever Told: Ecological Collapse, Progress, and Human Destiny', while we continue to take up the roles of explorers and pioneers, we must remember that caretaking and cultivation are also essential parts of being human. He says, 'If being human means anything, it means being, not just from Earth, but also of Earth, a part of it.'[6] In a sense, trying to escape this planet is only an act of trying to escape ourselves, and no matter how hard we try, that will never come to pass. We must ask: To what extent do these nihilistic, unreflective narratives of our relationship to the Earth exacerbate our disconnection to the planet and what alternatives exist to these stories? The question is simple – as posed by Margaret Mapondera, Trusha Reddy and Samantha Hargreaves in their essay exploring African ecofeminist alternatives to development, inspired by Arundhati Roy: If another world is possible, then who is doing the imagining?[7] Afrofuturism hands over this privilege of imagination to Africans and the African diaspora. Africanfuturism, a term coined by Nnedi Okorafor, is similar to Afrofuturism but is more explicitly rooted in African culture, history and mythology, and steers away from centring the West.[8] Both practices show us how African and Black communities can think about what *is*, what *can* and what *will* be. Afro- and Africanfuturism present to us a way of repairing the bonds broken by colonialism, slavery and other forms of oppression by imagining a plethora of alternative futures and working to make them possible in big and small ways – an infinite possibility of futures. And these stories, they are not only for Black people. The history of slavery and colonialism or other experiences of oppression are not unique to marginalised communities. They are our *collective* stories; they are the histories of the white community too. Our collective engagement with these stories teaches us how to see the 'end of the world' through the eyes of communities for whom the climate crisis is all but another ending.

To build a natural connection through a cultivated imagination is to realise that we have the power to change, and to be changed by, the world around us. Nowhere is this lesson more clear than in the opening words to the well-loved (and almost prophetic) Afrofuturist novel *Parable of the Sower* by the African American author and visionary Octavia E. Butler. She tells us that all that we touch, we change – and all that we change, changes us. When we are able to imagine, and then truly accept, our interconnectedness to all things on this Earth, we also realise our power.

For the benefit of those who have not yet read this marvel of a book, which I urge you to do, I'll give a brief synopsis. First published in 1993, the story follows Lauren Olamina, a fifteen-year-old Black girl who lives with the condition hyperempathy, which causes her to feel the pain of all those she can see, making life in the dystopian California she lives in unbearably overwhelming. She is surrounded by the constant drama and tragedy of fires, greed-driven and oppressive politicians, poverty and violence. Whilst the people around her wait desperately, and in vain, for their political leaders to do something, Lauren immerses herself in books on survival, horticulture, cultivation and land stewardship. Through her reading and her dreams, she steadily writes *Earthseed: The Books of the Living*, a sort of bible from which the opening words of this section originate. Undeterred by the bleakness that surrounds her, Lauren hatches a plan to create a better future for herself and all those willing to follow her – the Earthseed community. She sets off, leaving her imploding neighbourhood, in search of a suitable place to begin this community, building a diverse, interracial motley crew around her; her new family.

Lauren is someone who while living on the frontlines of environmental and social destruction, builds a vision so powerful, so compelling that she can do nothing else but strive for and journey towards it. She begins and continues on this journey not unscathed but neither deterred by the chaos, violence and fear that unleashes around her. She accepts that her community are unable to imagine a future outside of that which has been prescribed to them by politicians; but she does not lose sight of her knowledge that in order to shape the world, we must act. She tells us that all we touch we change and that all that changes, changes us. That we are not mere passive beings the world happens to. No, we have agency, we change the world as much as it changes us. We shape the future as much as it will, surely, shape us. Butler implores us to, like Lauren, feel deeply, be soft and strong in equal measure, be open, be brave and march ever forward to the images we have for the future. She asks us to accept that the only thing we can be sure of is change. Although some level of climate change–induced chaos is already expected from our present lack of inaction, this is not the only change waiting for us in the distant future. For me, *Parable of the Sower* is a mirror. Butler shows us that through our inaction – and even in the actions we desperately hope will impact our lives in a minimal way, one-shot, high-tech solutions – we reveal that maybe our approach to avoiding the end of the world is as much a strategy to avoid any change at all. But we must question who it serves to resist change. What magic might arise when we accept that the worst parts of this world must and will end, and make way for new realities to come into being?

Unlike most climate stories told in the science fiction genre, Butler's novel doesn't follow the common trope of counting on a billionaire-backed goal of abandoning the Earth and leaving for another planet or escaping the harshness of a future Earth in a high-tech vessel. Instead, Butler imagines a future based on the regeneration of this land and on bringing people together, facing our mistakes and building new systems. It's not to say that dystopian novels or films that centre technology or that present less than perfect visions of the future aren't entertaining, interesting or necessarily terrifying to watch – imploring us to do something. It's not that we must only tell stories of utopia – Butler herself was strongly and publicly opposed to the idea of building utopias.[9] Instead what she gives us is a reflection of how we, as a society, overcome the dominant narrative that tells us things will never change, and become open to stories that give us a 'long-term purpose that is "complex enough" to transcend [our] own individuality and bond [us] as a species'.[10] *Parable of the Sower* is not about imagining utopia, it's not about imagining the world saved by a single person or a single technology. It's about amplifying the power of community, collective vision, care for the Earth and how they might uncover a more generative and loving path to the future. As the professor Gerry Canavan notes, what Butler urges us to imagine is a world that is 'not perfect, not even perfectible, just *better*'.

12 From Imagination to Innovation

This section, IMAGINATION, has been an exploration of how we can and must resist the logics of colonialism and capitalism, which through a language of limitation trick us into believing that there exist no alternatives to the present world order. That the status quo of extraction, oppression and destruction are unavoidable and that a future built on mutual love, care and respect for the world – as nice a thought as it is – is impossible. Yet, through the stories of previously colonised peoples who have faced the impossible, we find that these ideas are not hard truths, they can and must be dismantled. By building a strong collective imagination, by dreaming and scheming together, the impossible can be endured, scaled, shaped and shattered time and time again. In the previous section, RAGE, we learned that our anger must be transformed into action, and an essential part of that practice of transformation lies in seriously, intentionally, nurturing our capacity to imagine better worlds. Our RAGE allows us to identify and respond to heartbreaking occurrences of exploitation and damage, to campaign and fight against them. But our IMAGINATION sets us firmly on the path from destruction to liberation, showing us the ways we might show up, the spaces we might create and the connections we must make with each other and the living world. We have seen that to radically change the world, we must also reimagine our relationship with it, coming to acknowledge and appreciate the life, the animacy, that lies all around us.

In the next section, INNOVATION, we go further,

witnessing how we build our visions for and connections to the living world into the solutions that we devise. In short, we will learn how it is that we bring our imaginations to life. How they become manifested in the rooted innovations we will read about in the coming chapters that materially address environmental and social challenges across the world. We will redefine what INNOVATION means beyond a Western industrialist view, beyond silver bullets and panacean solutions, in a rooted approach to building the worlds of the future.

Innovation

13 (How) Will Technology Save the Planet?

In the June of 2023, Apple released their first spatial computer, the Apple Vision Pro, a piece of technology that could have been pulled straight out of a *Black Mirror* episode. This self-proclaimed revolutionary 'spatial computer' allows 'users' to blend their physical world with the digital, bringing their overflowing emails and overwhelming notifications from their screens and into your physical space. The idea is that through the computer – fashioned as a pair of oversized, futuristic, mixed-reality goggles – we can stay present and connected to others as well as the natural world. Apart from a few sweeping drone shots of oceans and forests, it wasn't incredibly clear during the launch where the connection to the natural world fit into the experience. While through their installations technologists like Marshmallow Laser Feast (MLF), who we met in the IMAGINATION section, call us into wonder, awe, connection and embodiment of the living world, Apple asks us to be satisfied with a few high-definition videos of nature as we spend ever more time in their digital universe. I often wonder, as you might have, how the cultural psyche, and more importantly global action, of humans is affected when we are encouraged to engage superficially with a virtual nature instead of repairing the broken bonds we have with real-life ecosystems? What parallels can be found between us escaping to a flattened virtual nature and a billionaire attempting to escape Earth and move life to Mars? How does this strain of 'innovation' exasperate our already dwindling meaningful interactions with the rest of the living world? Many of us are worried about the domination of technology in our modern world and increasingly for its effect on global climate and biodiversity action. The lack of understanding or acknowledgement that the current climate and biodiversity crises are not just crises of measurement and calculation, but also crises of culture and connection to the ecosystems that we are a part of and that sustain us, feels unbearable.

The stories we are told about technology and its role in environmental action are ephemeral, ever changing and polarising. On the one hand, technology is hailed as a quick and magical solution that will take away all our environmental worries. Technology comes to the rescue of uncomfortable climate conversations full of fear and angst, and gives a reassuring pat on the shoulders of those unable to engage with questions about the systems of oppression and destruction at the heart of environmental breakdown. Often these narratives are pushed by governments, the private sector and research institutions that are inspired by the power of satellite technology and machine learning to detect deforestation and calculate carbon emissions from the sky. They are impressed by the rapidly falling cost of renewable energy technologies and have faith that the technologies of the future, like carbon capture and storage, will save us from excess emissions and dangerous average global temperature increases. And these beliefs have materially shifted the way we approach environmental action, directing large swathes of investment into the green tech space and shaping policy.

The 2022 Conference of the Parties of the UNFCCC (COP27), hosted in Sharm el Sheikh, Egypt, was described by the UN as a conference 'set to ratchet up the scale and effectiveness of innovation in tackling climate change' and there was an entire hub, the Global Innovation Hub, dedicated to this objective.[1] What was made clear in the UN articles on innovation ahead of the conference was the governmental and institutional lack of willingness or motivation to attend to the fundamental causes of the crises that face us – namely our obsession with and addiction to fossil fuels – with a preference for large-scale, industrial and resource-intensive technologies that allow business to unfold as usual. Many of these large conferences like the COP27 are inaccessible to the public and often seem like wastes of time, an opportunity for the elite and powerful to gather and propagate their own agendas or share empty platitudes and promises in a way that seems disconnected from every day. Although this may be partially true, the announcements, launches and intentions communicated in and around these events lead to tangible and undeniable changes on the ground, catalsyed by investment and backing for various innovation projects. Take green hydrogen for example, a substance that garnered much attention at COP27. Although the substance is produced using fossil fuel, it is renowned for its potential as an alternative energy source, making the production of green hydrogen an important goal in energy innovation spaces. It has come to be seen as an almost magical solution, which, if realised, will slash carbon emissions. In the lead up to COP27, a whopping $40 billion was invested by the Egyptian government in Egypt's green hydrogen industry, setting in motion an additional surge of major private investment in the country, totalling more than $100 billion, from energy giants in Australia, Norway and the UK.[2, 3] While other initiatives and campaigns such as the Loss and Damage Fund – a pot of money intended as compensation and reparations delivered from the wealthiest nations to other countries most affected yet least responsible for the climate crisis – struggle year after year for the lack of financial commitment, technological innovation to tackle the climate crisis seems to inspire a deluge of generosity and abundance from governments and private investors. History was made at COP27 with a Loss and Damage Fund being agreed on, but till the time of writing, no concrete numbers regarding funds have been detailed and much work lies ahead to get the money to the people who need it most as they suffer because of climate disaster and environmental breakdown.

In an article co-authored by Andrew Steer, president, and Kelly Levin, chief data scientist, of the Bezos Earth Fund, innovation was touted as 'critical' to achieving global climate targets, and the important role of the private sector in supporting governments to finance future innovation projects such as carbon capture and storage, generating renewable energy and producing electric vehicles was outlined.[4] In the first half of 2022 alone, private sector investment in climate technology start-ups reached $19 billion.[5]

The push from both public and private sectors towards innovation has, unsurprisingly, become enmeshed with the goals and direction of research. Recently in 2021, a programme at the Cambridge Centre for Carbon Credits (4C), University of Cambridge, was launched as a 'nature technology' project to halt deforestation through carbon credits with the idea that 'what you can measure, you can manage'.[6] The project hopes to tap into the $2 billion industry of voluntary carbon market, presenting a 'trusted, decentralised marketplace where purchasers of carbon credits can confidently and directly fund trusted nature-based projects'.[7] Generally, universities are attracting and committing to large amounts of investment in nature technology, seen as a financially sound decision that is at the same time thought to be the best choice for the planet.[8] And the ambitions of institutions to innovate in this field match the huge investments. In the UK, for example, 2021 saw the development of the Innovation Centre for Applied Sustainable Technologies (iCAST), a £17 million collaboration between the Universities of Oxford and Bath, and in 2022 Hull University secured £86 million of green funding from private investors, arranged by Lloyds Bank, to develop 'sustainable facilities and infrastructure'.[9] Investments are also being made outside of university initiatives: Private companies are ready to inject over £1.4 billion in tech start-up companies, many of which focus on green technologies, based on research by university students.[10]

But how have we come to decide that governments, well-resourced research institutions or billionaires are the gatekeepers or leaders of innovation? A recent study by the University of Sussex Business School revealed that 80 per cent of the investment in climate research and technology development over the last 30 years was spent across the UK, US and EU, with the UK dominating and accounting for 40 per cent of those investments.[11] What implications does this domination of innovation, originating primarily from industrialised Western countries, have on both the Earth and its people? What histories, cultures and alternative practices are being ignored? For many, technology exists as a crisis in itself, one that looms alongside, not counter to, climate breakdown. We are appalled by the level of emissions from technology itself, knowing that some large language models, like Google's BERT or ChatGPT, can emit as much carbon in one run, one cycle of learning for the algorithm, nearly a ton, as a round-trip flight between New York and San Francisco.[12] This statistic becomes even more staggering when you consider that AI models are 'run' many times during the training process. For large language models this could be as many as millions or billions of times! We are concerned that green technologies will inflict harms on marginalised communities, for example, the prevalence of surveillance using conservation technology where camera traps are used to spy on local forest communities. We are concerned by the greenwashing of mining companies like Rio Tinto, which announce to their shareholders and the world the need for their lithium mining in countries like Serbia. Their continued extraction has caused outrage from the Serbian community, with thousands taking to the streets in the summer of 2024 and continued research and resistance from researchers and activists including my friend and activist-academic at Cambridge, Sofija Stefanovic, who works with a coalition of groups resisting Rio Tinto's lithium project in western Serbia.[13] We are concerned about the heightened environmental destruction and human exploitation that could arise from increased mining of rare earth minerals like cobalt, used to power the batteries of electric vehicles, 70 per cent of which is mined in the Democratic Republic of Congo which continues to suffer violence, conflict and human rights violations. Some of us have heard of the plight of African forest communities battling to keep their lands from prospectors desperate to jump on the carbon offset train who are buying up forested land to sell for carbon credits. We worry that over-enthusiasm about sustainable tech will encourage 'climate delay', making investment pour into technological distractions that fail to attend to the social, cultural and economic roots of the climate and environmental crises that lie in disconnection, colonisation, capitalism, exploitation and overconsumption.

When it comes to technology, we live in a world of polarised views: Some are convinced that it will save the world, but many are sure it will not. In many ways, both perspectives are fundamentally flawed. We forget that *nothing* can save us. The American writer Rebecca Solnit speaks on this eloquently in her book *Hope in the Dark*, illustrating the incongruency of that word 'save' with what we aim to do in environmental and social movements. The word 'save', to her, implies that the world is tidy and final, that there will be some point in time when we will be able to exist in our happily ever after. Here, I'd add that to me, as a descendant of colonised communities whose oppression was justified through white saviourism – the idea that white people had a duty to liberate and lift Black, Native American, Indian, Micronesian and Melanesian communities, among others, from their 'degraded' positions in life – the connotations of 'saving' anything are unsettling. Debates about whether technology will save us or be the source of our demise neglect the fact that technology is neither neutral – politically, socially and environmentally – nor is it just about chatbots or electric cars. It is essential that we incorporate a more wide-ranging view of innovation, one that is not restricted to operating within the bounds of Western economies, politics and philosophies. For Kalpana Arias, eco-futurist and founder of Nowadays on Earth, the 'push and pull relationship towards nature and technology is because we're not fully embracing either'. We are failing to see the technological in nature and nature in the technological, unable to see past the 'techno-capitalist narcissistic perspective of what technology or innovation is'. Instead of considering innovations based on working with nature, such as the controlled burns used to shape the landscape by stimulating the growth of sequoia trees we saw in Chapter 8, we tend to think of technological advancement as purely human-made. Apple Vision Pros fit neatly into our perceptions of what technology and innovation look like. The images of technology we conjure in our minds are of silky, silver surfaces of metal, ever shrinking microchips, cleanly dressed scientists sat in laboratories, staring quizzically at large computer screens or diving headfirst into a tangle of wires and electronic components. We think of innovators as the people who bring us robots, cloud computing, driverless cars and even hologram concerts. Today, our understanding of innovation is closely associated to the commercialisation of technological inventions, perceptions that have been moulded since the dawn of the Industrial Revolution.[14]

The Industrial Revolution reshaped nearly every industry that existed at the time as well as created a plethora of new ones. In Britain, especially, the industrial period was heralded as an indispensable boon, bringing wealth and economic development – of course, at the exclusion of the new working-class labour force.[15] But industrialisation's influence on innovation was not always generative. Widely accepted as a significant catalyst of ecological collapse, the Industrial Revolution led to intense technological innovation and transformation, which has resulted in unspeakable loss and degradation.[16] In fact, industrialisation, fuelled in part by the wealth generated during the Atlantic slave trade, can be seen as a mere extension of colonialism – during the Industrial Revolution communities across the tropics were disconnected from their languages, cultures, land and, ultimately, their Indigenous technologies and innovations.[17] Even though the success of most colonial administrations depended on help and teachings derived from local ecological and technological knowledge, Indigenous peoples were characterised as 'backward' or 'anti-technological'.[18] These characterisations were used to justify European control over colonised lands. This violence was not only physical but intellectual. Through enforced colonial education, Indigenous communities were severed from their traditional and technical skills and practices, and taught to defer to the 'superior' European approaches.[19] The intention of this regime was to restrict, and in many cases wipe out, local-led, Indigenous or traditional technological development. Further to this enforced binary between Western science and Indigenous knowledge, a consequence of the suppression of Indigenous technology is our warped conception of technology as something separate from the natural world. We see this idea propagated through the works of influential Western thinkers like Francis Bacon who maintained that "scientific knowledge means technological power over nature".[20] Through the colonial project and Western philosophy, we have come to believe that technology must stand in stark contrast to natural ecosystems, a divergence from, and possibly an act of rejection of, Indigenous technologies that, as we will come to see, are often inseparable from the living world.

For Kalpana, this separation and invalidation of other ways of engaging in practices of innovation lie at the root of the tensions around the role of technology in environmental action and also explain how the Western understanding of innovation often leads to violence and destruction. 'Deep down', she tells me, 'we know that that's not the full story', it's not the only truth. Part of telling new stories about technology 'is where culture comes in, because [in] a lot of different cultures, nature really guides the way that innovation takes place', the living world continues to exist as one of the greatest innovators and technologists on the planet. If innovation is to have an effective, equitable and sustained role in environmental action, we must reimagine and redefine it in the Western context; we must remember and reconnect with past as well as present forms of cultural and ecological technology.

When thinking about how we can redefine innovation, I was drawn to the root of the word itself. 'Innovate' comes from the Latin *innovare*, to make new but also to *r*enew, and 'technology' comes from the Greek for *art* or *craft*. The roots of these words guide us to a different view of the practices that must shape and serve not only sustainable, but regenerative, technological innovation. Not the practices of the rich and industrialised that serve to protect their own kind, bolster their egos and avoid the social and environmental dimensions of the environmental crises; but technological innovation that serves as a tool of creativity and craft. A tool that is informed by a multiplicity of influences, cultures and people, and that exists in harmony with the planet, makes the world anew. As the former Canadian Senator Murray Sinclair – whose Ojibway name, Mizanay Gheezhik, means 'the One Who Speaks of Pictures in the Sky' – shared beautifully at the 150th Indigenous Innovation Summit in Winnipeg, in so-called Canada: 'Innovation isn't always about creating new things. Innovation sometimes involves looking back to our old ways and bringing them forward to this new situation.'[21, 22] To renew the world isn't always about making new *things*, but allowing the past to help us see, think and live in new *ways*.

As is often observed, despite the sheer diversity of Indigenous thought, tradition and cultures across the globe, many synergies arise and it is heartening to see the parallels between Sinclair's beliefs on innovation and the lessons taught through the lessons of the Ghanaian proverb and tradition of Sankofa, symbolising a 'return to your roots', translating literally to 'it is not taboo to fetch what is at risk of being left behind'. It is a tradition that urges us, as we hurtle, march toward, or are dragged into the future, to treasure the knowledge of the past.

Working at the intersection of technology, ecology and justice myself, the concept of a rooted innovation is fundamental to me. My experience of this work is complex. I am fascinated and driven by the power of science and technology, of our ability to advance our knowledge of and connection to the planet, and to create, as if from thin air, revolutionary and life-changing inventions. But I also acknowledge and work to amplify the reality that technologies definitely do not appear out of thin air despite misleading phrases like 'the cloud', which may make us think that they do. From data centres to hardware to the electricity they use, technologies are material manifestations born from the Earth and from the labour of, commonly, underrepresented and vulnerable communities. Despite my love for and interest in technology, I also have a deep reverence for all things spiritual, traditional and cultural. I am an eager student of the living world and highly motivated by the fight for social and ecological justice, well aware that none of our modern-day technologies leave the Earth unscathed. Yet, neither am I an individual too fond of binaries: I believe that it is with the knowledge that there is no black and white when it comes to technology, that the potential for transformation and harm lies in a common, messy space, that gives rise to more scope for iteration, experimentation and innovation. Through my research, I am split across multiple ecosystems and exist in multiple forms. Not just between the obvious and starkly different environments of the University of Cambridge and a tiny rural forest community in Ghana, but also between different disciplines and schools of thought. The hyper-masculine, sometimes arrogant, scale-centric, fast pace of computer science; the steadfast critique, vigilance and advocacy of sociology; and the long-standing, slow(er)-moving, impact-driven space of conservation. While the core objective of my doctoral studies is the creation of a novel innovation, a piece of technology, that will allow conservationists and forest communities to analyse forest soundscapes to monitor biodiversity and health, the roots of this work is philosophical and methodological. It will present new ways of thinking about how we develop and use conservation technology; the role communities are able to play in not only designing that technology but in making decisions that best serve themselves and the environment from it. As my work unfolds, I often come back to the idea of rooted innovation, reflecting on the ways in which I am honouring the social and ecological, even as I spend hours behind a computer screen. In this way, rooted innovation is a process, a perspective, a way of thinking, not only reserved for technologists in the formal sense of the word, but also to be used in our every day. It's a way of looking more closely at the objects and structures around us and embedding them in the greater ecosystems of nature and culture. The phone in your hand is not just a device, it is the thicket of trees that was felled, the heaps of earth that were moved and the physical effort that was made to dredge that earth to mine the lithium within it.

A tree is no longer a tree, but an electricity generator, a life giver, and an embodiment of adaptation, migrating as our climate changes. The process of rooting in innovation requires us to be more attentive to the world around us, to be amazed by the traditional, 'simple' or ordinary. To see a bucket and a cup used in African households not only as a marker of poverty but as an eco-friendly, water-saving bath, or old clothes and rags as new 'zero-waste' mops. Innovation is not always about birthing a shiny new object but is invariably about having the ability to solve problems in the unlikeliest of places, with the resources available to us in ways that honour each other and our planet. It is a practice in slowing down enough to hear the reverberating wisdom of the past, to observe the ingenuity within the living world, and to listen to and be led by the experiences and needs of those most oppressed by both environmental breakdown and technologies of extraction. In this way, finding roots in innovation is not only about an applied, technical or manifested intelligence but, more importantly, emotional intelligence. We may be called to innovate because our community is suffering, or because we are afraid for the children in our lives and those that will come after them, or we might be completely infatuated and energised by the prospect of creating, as if by magic, tools or ways of thinking that can change the world in immeasurably positive ways. This chapter is a space to explore innovations, past and present, that have the potential to change the future for the better. It is a space to understand how we can contribute to and encourage a rooted innovation, grounding it in the teachings of the past, and at the same time acknowledging the opportunities for renewal, invention and problem-solving that defy our limited view of technology.

14 Wakandan Cosmology: A Blueprint for Rooted Innovation

In Chapter 11, we witnessed how Afrofuturist visions allow us to imagine liberated lives through literature. When thinking about how we defy limited views of technology and innovation, I am again drawn to the tradition of Afrofuturism, but this time to ideas brought to us on the big screen. The kingdom of Wakanda, the home of Black Panther, presents us with a vision of a society where the wisdom of the past, the beauty of the living world and the advancements offered by technology are elegantly married. And where the view that technological excellence resides only in the minds and institutions of the West is fiercely rejected.

In the Black Panther films, the countries of the West are determined to enter the hidden society of the Wakandans, intrigued by the superiority and power of their technological advancements. Although it is a story about current-day Africa that is often left untold, it is interesting to observe the brain drain that nearly every major technology company, from Google to Facebook, facilitates by setting up head offices across the continent, poaching exceptional African talent for their own development. Wakandan technology is not based on Western ideologies, it does not resemble the technologies developed in Silicon Valley – often seen as the global centre of technological development. Instead of exploiting natural resources for profit, Wakandans covet and harness their technology for collective empowerment; they do not export it, as is often the case, for capitalist gain that rarely serves the most vulnerable.

This approach is aligned with philosophies introduced to the continent by prominent figures such as Kwame Nkrumah. Nkrumah led the Gold Coast struggle for freedom from Britain and the creation of the new nation of Ghana, leading the country from the time of its independence until the military coup of 1966. He was also a founding member of the predecessor to the African Union, the Organisation for African Unity. It was at the opening of this organisation that he made his famous speech where he called for Pan-Africanism, stressing the necessity for Africa to 'unite or perish'.[1] A notable component of his ideology was 'an Africa, unified, modernized and developed with science and technology for the benefit of all Africans'. Reading Nkrumah's words and thoughts on technology gives me the chills – his abilities of foretelling and envisioning African technological futures years before Afrofuturism was even conceived is astonishing. In his 1967 article, *African Socialism Revisited*, he wrote that 'socialism in Africa introduces modern technology [that] is reconciled with human values, in which the advanced technical society is realised without the staggering social malefactions and deep schisms of capitalist industrial society'.[2]

Wakanda is but one realisation of such an 'advanced technical society'. One that, it is important to acknowledge, does not function without the strength, intelligence and leadership of women. From Okoye, the general of the Dora Milaje, Wakanda's women-only special forces, to Nakia, a force for nature in the Wakandan Intelligence Agency, women hold some of the most powerful, respected and trusted positions in Wakandan society. In the sequel, *Wakanda Forever*, we see the cross-cultural partnership of Shuri and Riri, the former the Wakandan princess (and later, Queen) and the latter a gifted nineteen-year-old

African American student at the prestigious Massachusetts Institute of Technology. Together, in Shuri's lab, they fuse their ingenuity and intelligence, and combine artificial intelligence, biology, engineering and material science to defend Wakanda. This contrasts with the reality that in Africa only 3 per cent of women in higher education are enrolled in information and communications technology (ICT) courses. In the US, Black women represent only 2.9 per cent of science, technology, engineering and mathematics (STEM) degree holders and in the UK the story is much the same.[3] In 2013, the British-Nigerian computer scientist Anne-Marie Imafidon founded Stemettes, an organisation inspiring, supporting and encouraging girls, young women and non-binary young people into technical fields, prompted by her experience of being one of the only three women studying maths and computer science at her university.

As a young Black woman working in technology, I find my own experiences of racism and sexism, alongside observations of encoded oppression, agree with many of the feminist perspectives that recognise that technology is not neutral, but rather shaped by the interests and values of its creators, which for the majority of history have been white men. And though it might be tempting here to conflate the experience of Black women in STEM with that of women more generally, it is important to note the unique experiences and harms borne by Black women in technological spaces. From being hypervisible in the workplace, held to impossible standards of excellence to prove their worth and 'keep their seat at the table', to experiencing microaggressions, racist slurs and the invalidation of their work, the struggles of the Black woman in technology continue long after she is 'generously' accepted into an institution or organisation. Black women's ideas are excluded from mainstream technological dialogues and instead they have to fight and risk their jobs to stand up for what they believe in, only for their works to be plagiarised by their own companies.

As described by Yolanda Rankin and Jakita Thomas in their paper 'The Intersectional Experiences of Black Women in Computing', 'The invisibility of the Black woman has been critical in maintaining social inequalities.'[4] Why? Because as the famous saying – originating in a collective statement presented by the Combahee River Collective, a group of Black feminists, in 1978 – goes, 'Until Black women are free, none of us will be free', and so suppressing Black women helps maintain systems of oppression and discrimination. The relevance of this fact in the world of technology cannot be understated. From Joy Buolamwini, who discovered the use of racist facial recognition systems in her PhD thesis at MIT, to Timnit Gebru, the former Google researcher who was fired from her position (Google says she resigned) for campaigning for transparency about the environmental and social harms of the company's large language models that was uncovered in her research, Black women have the ability to highlight a multiplicity of embedded and potential harms being caused by technology across environmental, feminist and marginalised perspectives, to list only a few.[5] The depiction of two, notably young, women using their scientific excellence for the protection of a technologically advanced African nation in Wakanda Forever deliberately challenges the idea that Black women are incapable of, unsuited to or unworthy of careers and leadership positions in STEM.

It is not only gender discrimination and racism in innovation that Wakandan life challenges. Unlike many

Western visions of technological progress, innovation in Wakanda does not put humans at odds with the living world but brings Wakandans closer to it. The portrayal of Wakanda's abundance in the valuable and potent natural resource vibranium is inspired by the wealth of natural resources that exist across Africa. However, although Wakanda is presented as a land never having been touched by colonialism, in the real world, African nations continue to be plundered. Vibranium, with its myriad protective, powerful and ritualistic powers, lies at the heart of Wakanda's economic, social and technological success. It is used to protect and enhance the lives of all Wakandans in what can be described as an autarky, a nation that is wholly self-sufficient, an extreme contrast with the majority of present-day African countries, which remain dependent on the export of natural resources for survival. From the Democratic Republic of Congo's coltan – 60 per cent of which is mined by thousands of vulnerable people, including an estimated 40,000 children and teenagers – which is indispensable for the development of nearly all modern technological devices, to Ghana's much sought-after gold, only 1.7 per cent of the profits of which are received by the country itself, Africa is the victim of theft and pillage by the international community.[6, 7] But in *Wakanda Forever*, a different reality is shown, one where Queen Ramonda, the leader of Wakanda in the film, gives a powerful speech to the member states of the UN, warning them of Wakanda's zero-tolerance policy of attempts to find, steal or create vibranium.

Much of the critical commentary about why vibranium is so diligently guarded points towards the Wakandans' need to remain technologically superior to the rest of the world. That without vibranium, the country loses its advantage in the global political sphere. Though this may be true, as an environmental advocate, I also notice some other influences. We do not see the Wakandans using their technology to oppress or subjugate other nations; in fact, they rarely engage in war, violence or combat unless in defence. I interpret the coveting of vibranium as an act rooted in the fundamental difference between the systems and worldviews of Wakanda and that of Western and developed countries. Vibranium is a material that is quite literally out of this world, a substance that disobeys the laws of physics. It renders Wakanda invisible to the rest of the world, absorbs the kinetic energy of any projectile and when weaved into fabric, alters clothing so that they become bullet proof. Yet, it is also the substance that powers the sacred *heart-shaped herb*; it is a plant unique to Wakanda, which, when ingested in ritual, transports the consumer to the ancestral plane where the oral tradition and wisdom of the elders is preserved and transmitted. As described in an essay by Ryan Kennedy entitled 'Society, Technology and Cosmology in *Black Panther*, vibranium is not only a substance of African technology, but also of African cosmology.[8] A cosmology can be described as the dominant narrative(s) that informs a people's ideas about the world. The cosmology of the Wakandans presents us with an exercise in remembering. Remembering traditions, beliefs and practices that, in their diversity and plurality, used to be (and continue in small pockets of the African continent to be) the foundation of African societies. For the Wakandans, vibranium is not just a resource to be exploited. It is not a means to cause destruction and harm neither is it an inanimate mineral. Vibranium is multifaceted: Yes, it provides technological advancement, but it is also a form of traditional medicine. Vibranium is a part of

Wakandan heritage, with the wisdom and sanctity of the ancestors imbued within it. Vibranium is a metaphor for the coexistence of technology and tradition.

We see this notion of harmony between technology, nature and culture manifest in other ways in Wakandan society, most notably in the architecture and design of the Golden City, Wakanda's capital. The city design was conceived by the African American production designer Hannah Beachler, who has worked on an impressive array of visual projects, including Beyonce's *Lemonade*. In 2019, she made history by becoming the first Black woman to win an Oscar for her production design on the first Black Panther movie. In the Golden City, towering skyscrapers emerge from the ground like ancient forest trees, interspersed with aerial gardens and dwarfed by the densely forested mountains in the distance. There are vibranium-powered driverless buses and an elevated, magnetically levitating high-speed rail that transports the city's residents in what is a public transit utopia – built in direct response to Beachler's observations of the barriers to fast, efficient and safe public transport for marginalised communities in America.[9] It took eighteen months to finalise the concept for Wakanda, with everything from the Wakandan alphabet to the best Wakandan tribal restaurants catalogued in a 500-page 'bible'.[10] The Wakanda bible is the product of in-depth research by Beachler, who studied the traditions of tribes across Africa, the practices of traditional Congolese fishermen and the science behind mining, landforms and Africa's ecology.[11] The Golden City is a thriving and advanced metropolis where technology serves the needs of the people rather than the other way around as is witnessed in many dystopian visions of the future. In contrast to the dominant approach of Western technology companies, which seek to turn people into 'users' and to extract profit from them with questionable positive impact on those very people or the planet, Beachler wanted to envision a city where the people were the most important thing. So instead of sprawling suburbs or gentrified towns that push local people and cultures away, the Golden City is abuzz with tradition. The skyscrapers exist in tandem with bustling markets, distinctive to many African cities, with stalls selling traditional foods and wares. Hovercrafts zoom over the pristine river that courses through the city as fishermen in traditional canoes wave up from below. Wakanda is the most technologically developed country in the world, yet it has conserved its natural resources with mountains, forests and waterfalls abundant in the region and has sustained its ritualistic and ceremonial traditions. These include traditions of inter-tribal harmony, or at the least, cooperation; a rendering of a true Pan-African movement that is yet to exist. The coexistence of high technology, deep tradition and ritual, and reverence for the living world in Wakanda is incongruent with Western and colonial conceptions of innovation and development. Conceptions that see traditions, folklore and sentimentality for nature as backwards and regressive. What I find incredibly poignant is Beachler's creation of a city and people that in her words, 'know everything about their past', and I would add, use and champion that knowledge rather than discard it. This is the essence of rooted innovation. Whether our past has been hidden or stripped away from us as oppressed people, whether we bury our heads in the sand for fear of the horrific past of our ancestors in which they inflicted injustice on others, without reconnecting to, reckoning with and rooting in the past, innovation will continue to be a tool of oppression and persecution. Wakanda is created through the American gaze, but it is an Afrofuturist vision of a sustainable, technologically advanced society that despite being imperfect is a blueprint for a future founded in innovation that is regenerative, that sustains life and preserves culture.

When thinking about centring indigeneity and tradition in innovation, what inspires and motivates me are the places, like Wakanda, where the art of the past and the ingenuity of the present coalesce. They are akin to the confluence of a newly born tributary and the parent river, simultaneously holding immense histories and carving out a multiplicity of futures. Even as our eyes and minds are drawn to the dominant, polarising conversations on technology, communities, researchers and engineers around the world are already creating worlds where the future is firmly rooted in the past, honouring our need for cultural, ecological and technological innovation. A research team from Brunel University London is spearheading efforts to document, distribute and amplify African Indigenous knowledge by co-designing a large-scale database of Indigenous African practices which contribute to the food production, preservation and consumption needs of over 80 per cent of African citizens. The work they have undertaken is not only about generation of new ideas, tools and technologies for ecologically sound futures but preservation too. Preservations of knowledge systems threatened by migration to urban centres as well as land grabbing by and the domination of multinational companies in rural African regions.[12]

One of the most inspiring examples of the many projects conducted to centre African Indigenous knowledge in environmental innovation is the work of Muthoni

Masinde. Working with small-scale farmers in two Indigenous Kenyan communities – the Mbeere and the Abanyole – Muthoni Masinde is the scientist who brought the Wakandan blueprint to life through her doctoral research. Growing up in a small-scale Kenyan farming community herself, Muthoni was keenly aware of the struggles brought on by the worsening climate crisis on her family and neighbours. Agriculture contributes over 50 per cent of Kenya's GDP, and 90 per cent of the food grown is entirely dependent on the country's two wet seasons, the long rains that occur between March and May and the short rains that occur from October to December.[13] Climate change is severely impacting the rain patterns, resulting in reduced rainfall in the long season, intensifying the ongoing drought in the region.[14] For the 7.5 million small-scale farmers producing 80 per cent of the country's crops, the lack of rain spells disaster, impacting not only their income but their personal food security.[15] Meteorologists have attempted to support farmers by providing drought prediction systems that allow for sufficient planning and guide the timings for farmers to sow seeds. Yet, these systems make predictions on too large a geographical scale which prove to be less than useful at the local level; the information also fails to reach enough people as a result of poor dissemination tools. These are common problems of modern environmental technologies that aim to support marginalised communities. They are designed and developed, though with good intention, far away from the intended beneficiaries, with no grounding in their needs, ideas, lived experience or knowledge. To remedy this situation, motivated by the needs of her people and her passion for technological innovation, Muthoni combined her experiences of growing up in rural Kenya and her skills as a computer scientist to create ITIKI – the Information Technology and Indigenous Knowledge with Intelligence – a tool for drought prediction. She knew that 'small-scale farmers in Africa have always based their major decisions on Indigenous knowledge of weather and climate patterns'.[16] In her thesis she described how Kenyan farmers are informed about oncoming drought by the living world, observing 'lunar cycles, the shape or position of the moon and the patterns of stars' as well as the 'the behaviour of animals and birds and the appearance of plants'. As with modern-day meteorology, Indigenous methods also include the study of 'air, temperature, cloud colour and the direction of the wind'. The rainmakers of the Abasiekwe clan of the Abanyole community are well known for the utilisation of their Indigenous knowledge to summon water from the heavens and have a history of providing local and trusted drought predictions to the people of Bunyore and beyond. Unsurprisingly, these practices were largely left out of modern scientific approaches to drought prediction, deemed 'too primitive' and 'imprecise'. Refusing to wait for the West to 'discover' and exploit the knowledge of her people, Muthoni set off, determined to rewrite the norm and displace the rigidity found in formal climate science by integrating her community's traditional technological knowledge with cutting edge machine learning prediction techniques in what she calls 'a mutual symbiosis' between the two methods. By developing speech-to-text and text-to-speech mobile applications, Muthoni was able to build a database of Indigenous drought predictors from the Mbeere and Abanyole communities, allowing them to describe in detail their practices and techniques. She then set up a network of hundreds of wireless weather meters and sensors calibrated against conventional weather stations to provide more granular meteorological data for her system. Then came the machine learning to grapple with the vast volumes of data being collected and to generate accurate predictions of oncoming drought based on Indigenous knowledge and real-world weather data.

ITIKI is now deployed across Kenya, South Africa and Mozambique, supporting over 10,000 farmers and 64,000 hectares of farmland. Using a large sensor network, recording climate and weather patterns in the field sites, and machine learning algorithms, trained and embedded with 'holistic Indigenous knowledge' drought markers, the tool has succeeded in providing useful and actionable insights for social and environmental benefit.[17] In conversation at SciFest Africa 2021, Muthoni made clear the motivation grounding her approach: 'If we stick to the core science . . . without thinking about the people whom we are creating the products for, we will then fail to create an impact.'[18] She went on to make clear that 'contextualised innovations built by, with and for local people, have a higher chance of succeeding and Indigenous Knowledge Systems bridge this gap because they support ways that are culturally appropriate and locally relevant to them'. Learning from Indigenous technology or redefining innovation through the lens of indigeneity has more to do with how we think about technological development, the philosophies that inspire technological solutions and the ethics they are developed and built on, rather than seeking the blueprint of Indigenous technologies for replication in unrelated contexts and locations. Innovations based on Indigenous cultures, traditions and actions must be led by and co-created with Indigenous communities to ensure the colonial practices are not perpetuated and to enable Indigenous resistance to colonising technologies and wider systems of oppression.[19] Muthoni's work teaches us that finding roots in innovation is about the ability to develop the tools of the future by honouring the physical, spiritual, traditional and ecological wisdom already available to us.

We often look up and beyond, building higher and bigger, reaching for the stars and at the same time looking forward, predicting and building for the future, or waiting for future generations to come up with technologies based on what we need today. But, as we will discover in the next chapter, sometimes innovation rises, literally, from roots put down centuries ago.

15 Roots of Innovation

I remember my first journey to the forests in Ghana. The Land Cruiser jolting and jumping between the wide potholes that scarred the red clay earth, flanked tightly by thickets of trees that stretched for miles. To the left, green. To the right, green. Up ahead, green. Greens of every shade and depth, verging on iridescence. It is this verdant domination that my imagination conjures up when I think about Meghalaya, a place I have had the privilege of experiencing only through the images, videos and words of others. The state of Meghalaya is located in northeast India, sharing borders with Assam, another Indian state, and Bangladesh. It is home to some of the world's richest biodiversity: a mosaic of undulating hills, deep gorges, dense forests and betel plantations.[1] Meghalaya means 'Adobe of Clouds' in Sanskrit, but one of the state's greatest wonders lies firmly on the ground. Amongst the valleys of the Khasi (and Jaintia) hills can be found almost one hundred living root bridges.[2, 3] I could stare at images of these structures for hours – intricate, delicate, strong and imposing all at once. The living root bridges, known locally as *jingkieng dieng jri*, are an embodiment of the intertwined relationship of the Khasi people with the forest and date back as far as 100 BCE.[4] Ranging from 15 to 250 feet, the bridges take between fifteen and thirty years to construct from the aerial roots of the *Ficus elastica* (Indian rubber tree).[5] A branch of this tree is usually planted on the bank of a river or canyon, with the young aerial roots trained, often across a bamboo framework. These roots produce more roots which are trained in the same way and the process repeats, until the roots can be planted on the bank opposite where they originated.[6] What is created through the tangle of branches and roots is a magnificent structure. Astonishingly, without a single nut, bolt or piece of scaffolding in sight, these bridges increase in strength over time, becoming more resilient to environmental stress and strain, able to bear the weight of up to 50 people at a time.[7] As the Nigerian scholar Uzoma Samuel Osuala describes, Indigenous technology is the 'art of doing things', echoing the origins of the word 'technology' itself. They are active practices that embody craft, culture and connection. The living root bridges embody not only craft but also function. The bridges are a piece of 'critical' infrastructure that enable community members to navigate hills and cross over rivers to trade and connect with others, and to reach their agricultural lands; they are integral to Khasi life.

Over the last decade, Meghalayan activists and land defenders have experienced ongoing violence, harassment, disturbance and destruction as hydroelectric dam projects are established, which displace communities and destroy biodiversity, or as a result of the continued deforestation to carry on illegal mining of coke, which is used to make steel.[8, 9]

The destruction wrought by both heavy industry and the intensifying monsoons in the region have not only decimated many of the bridges but also, more importantly, destroyed the Meghalayan people's connection to the practice of building root structures. The Khasi origin story tells of the god U Blei who, angered by the greed and selfishness of his people who had forgotten their commitment to respect their connection with their ancestors, sent down the most monstrous tree, the Tree of Doom, Diengiei. The tree was an oak that grew so big that it cast a shadow over all of the Earth. It is said that U Iakjakor, a Satan-like figure, sometimes described as a bird, convinced humans to cut down the giant Diengiei tree, severing completely their connection to the living world and their god.[10] The lesson from this story, it is believed, is that when humans separate themselves from 'nature's essence', they lose their ecological conscience.[11] The disconnection of tradition and innovation was a deliberate one, and if we are to learn from our history, it is up to us to help repair or facilitate the reparation of those broken bonds. One individual, Morningstar Khongthaw, is working tirelessly to reconnect his community, and the world, to the roots of Meghalaya. In 2018, he founded the Living Root Foundation, an organisation reviving the practice of building and maintaining root structures in villages across Meghalaya. Speaking to him over Zoom, he described that his work is about protecting the cultural heritage of the Meghalayan people and about advocating for the 'communities who have been neglected because of the emergence of modern innovation, so called development that is more interested in concrete, cable and steel' than in engaging with rooted narratives about innovation. As Julia Watson, designer and author of *Lo-TEK*, a book documenting nature-based technologies for climate resilience, told me, Indigenous innovation is the act of being 'principally engaged with oral transmission of knowledge, [where] narrative is the form of knowledge'. Storytelling as an approach to innovation is often snubbed by Western science and seldom implemented in the design of new technologies; but for Morningstar, stories are the key to preserving the majesty and utility of the living root bridges. Much of Morningstar's work involves liaising with and listening to the elders of the villages he visits.

Reflecting on the importance of intergenerationality when it comes to innovation, he tells me that 'we need to meet those elders, we need to meet those people who still have the connections'. In Western cultures, we have become greatly divorced from our elders and we discard the wisdom they hold as outdated and backward; but for Morningstar, as for many other Indigenous communities, the knowledge held by the elders is the key to the future. Contrary to popular conceptions, Indigenous or traditional innovations are not specimens frozen in time. They are dynamic, constantly evolving and firmly exist in the present. As Morningstar tells me, the practice of the elders is a practice in imagination and visioning. They do not design and build bridges for their present but for their descendants twenty-five years in the future – 'how they build is how they imagine the future'. Morningstar is working to not only preserve this Indigenous wisdom but build on and modify it to create new sustainable structures that will extend the longevity of root structures. Whereas these structures would have previously needed maintenance after three to five years, new processes have extended the bamboo lifespan to ten, even fifteen years. He is currently working on new pieces of architecture, such as shelters and tunnels, that can support communities in the forests and in the growing urban centres. At its core, he says, 'It's about coalescing teachings of nature and elders or old ways with the power and advancements of new technology.'

The living root bridges, like many other Indigenous innovations, have inspired a whole generation of designers, architects, engineers and technologists. In 2005, a team of German architects led by Ferdinand Ludwig built the 'Baubotanik Footbridge' as an experimental building inspired by Ludwig's travels and research in Meghalaya. The structure was erected growing vertical and diagonal bundles of willow branches that are great at rooting themselves in soil. The footbridge itself was an arrangement of floating steel, held up only by the willow bundles.

Whether built by Indigenous communities or inspired by Indigenous principles, building with and for the benefit of nature is known as *regenerative design*. Defined by the Royal Institute of British Architects (RIBA), regenerative design is 'a system of technologies and strategies, based on an understanding of the inner working of ecosystems that . . . regenerate rather than deplete underlying life support systems and resources within socioecological wholes'.[12] It is a way of combining the needs of society with the wisdom and sustenance of the living world. Regenerative design manifests in various ways, but often as *biomimicry*, 'nature inspired innovation'. This is a practice documented in many cultures around the world and can be observed even in Greek myths such as that of Daedalus and Icarus, who mimicked the design of bird wings in order to fly, with varying degrees of success, towards the sun. Biomimicry has become a buzzword in the technology and design spaces, and we are seeing a flood of new ideas that look to nature for inspiration and adapt traditional practices in exciting and life-sustaining ways. From the creation of Velcro by George de Mestral who studied the fibres of burrs after being exceedingly annoyed at their capacity to grasp onto his clothes on walks, to the building of apartment blocks that rise from the ground as self-contained forests, like Stefano Boeri's 'Vertical Forest' in Milan. Some applications are dizzyingly high-tech, such as the work of Natsai Audrey Chieza, founder of Faber Futures, who created zero-waste fabric dyes made from pigment-producing bacteria that require 500 times less water than traditional dyeing. Others are stunningly simple, like the practice of manufacturing textiles from wheatgrass as the bio-designer Zena Holloway does.

Speaking with Zena, I had the opportunity to learn more about her self-titled *rootfull* design practice. Originally trained as an underwater photographer, Zena always had an interest in the intricacies of the living world, and especially in the grassy lifeforms that sustain underwater ecosystems. One day, when she was walking by a local river, Zena noticed some roots. Her head had been full of thoughts of mycelium, of the patterns and shapes they could make and whether there was an opportunity to innovate the fabric and materials space with them. But passing by these roots set her mind on a completely different path. Not knowing where she would end up or how it would go but motivated by her frustration at the waste and pollution generated by the fashion industry as a consumer, she began to experiment. Setting up shop in her garage, she began to grow seeds everywhere. Seeds of all types scattered in pots and bags. She admitted that her friends and family thought she was mad, and a bit obsessed, but her vision would soon pay off. Her experimentation brought her to wheatgrass. It didn't take long to grow – around two weeks – and didn't require any soil. In that time, the seeds could sprout roots of up to 20 centimetres. She cut off the grass tops and fed them to her chickens and got started on designing textiles using the roots. Just as the living root bridges were trained around bamboo, Zena's root material is trained on beeswax. Explaining the process in more detail, she tells me that 'once the seed is sown, it is grown across beeswax templates ... looking for water, the roots grow down into the templates and run through it [as like in] a maze'. Once the main stretch of material has been created, it is formed, shaped and moulded, and after twenty-four hours, the roots become stiff, holding their shape. What emerges is an incredible collection of forms and objects, from dresses to lampshades and plant holders, all made from two simple ingredients, water and wheatgrass seed. Zena has had the honour of exhibiting her designs across the UK, including at the RHS Chelsea Flower Show. Reflecting on people's responses to the designs, Zena remarks that so many people are surprised that they are made of roots. Surprised that a material so simple could look so beautiful. To Julia Watson, as she tells me, this is the epitome of successful innovation, where it 'is so implicit within an environment and a culture and an understanding of a human relationship with their natural environment ... that to the outward view, it almost dissipates, it's almost invisible'.

This invisibility is something Zena honours, but she also feels that it is important to highlight it. She is reminded of journeys driving past fields of crops, and how, she says, we rarely 'think of all those root structures that are buried in the soils that we never see, you only ever see half the plant, you know, and it's about this textile, a living textile that's buried in the ground that we never see. And interestingly, all I'm doing is lifting it out of the ground'. Zena's practice is one that shows us that innovation can mean something as simple as revealing the innate creativity and technology embedded in the living world. Exhibiting its beauty to not only inspire but to build alternative systems so well designed that their roots are imperceptible. In our next example, innovation comes, not from roots, but from the Earth itself.

16 The Future Lies in the Clay: Iran's Energy-Free Cooling Solutions

It was getting hotter ... people were outside buildings, clustered in doorways ... round-eyed with distress and fear, red-eyed from the heat ...

These lines are from the opening chapter of Kim Stanley Robinson's epic climate novel *The Ministry for the Future*. The novel starts in India, in the midst of a deadly heatwave. The conditions were dire, the electricity had cut off, meaning the few buildings blessed with air conditioners had become death traps. In desperation, people dashed to a nearby lake but were met with danger rather than respite. The water had been mercilessly heated by the sun, way above body temperature; many were dead. People had nowhere to go, no water to drink, no rescue services on the way. They could but endure.

By 2030, India is set to be the country most threatened by intensifying heat levels, with more than 770 million people expected to live through 'highly dangerous' heat conditions at least two weeks out of the year.[1] With the number of days a year exceeding liveable heat levels soaring, more people face the threat of health problems, such as heat stroke and, in the worst cases, death. As I edit this section in May 2024, parts of India have already experienced temperatures as high as nearly 53° Celsius.[2] Without innovation, providing the necessary energy to cool homes and businesses in South East Asia will be incredibly carbon-intensive with a predicted 23.1 million tonnes of carbon expected to be emitted. With electricity supply being affected by outages during heatwaves, it is becoming an increasingly vulnerable source of energy; Indian buildings are in desperate need of alternative air-conditioning systems to provide respite and safety. Ant Studio, a Delhi-based design company founded by Monish Siripurapu, is trying to come up with just that. Taking inspiration from our pollinator heroes, bees, and built literally from the Earth, they have created a 'zero energy air conditioner' called the 'CoolAnt'.[3] Inspired by the design of a beehive, the cooling system is built from individually cast terracotta cones, formed using traditional methods, and assembled into a large modular circle. Siripurapu's work was motivated by his first-hand experience of India's blistering heat and was catalysed by a collaboration – working with a factory owner whose large diesel generator was flooding the building with not only pollution but also unbearable excess heat, posing health risks to his workers. Inspired by ancient Indian techniques that use water and clay to lower temperatures, the CoolAnt's design passes recycled water through the terracotta cones.[4] When air passes through the cones, the water evaporates providing a cooling effect. Siripurapu's design not only provides safer working and living conditions for the people of India, but through experimentation, he has stumbled across a bonus innovation in the previously unwanted growth of moss that occurred when the system was used outside. Originally seen as a nuisance, he found that the moss could act as an air purifier if allowed to grow on the clay cones. With extreme heat causing heightened air pollution, this extra innovation provides an additional layer of safety.[5] Siripurapu's aim is to create smaller arrays of his terracotta beehive to supply cooling to cafes, train stations and public buildings.

The technique of using clay, air and water for energy-free cooling is an ancient and widespread practice that can also be found in other cultures, such as in windcatchers, *bâdgirs*, of Iran. The city of Yazd is home to the largest number of these ingenious clay structures, hailed as 'ancient engineering marvels', that soar above the rooftops.[6] Bâdgirs are said to have originated in ancient Egypt but became prolific and indispensable cooling systems across Iran, including Yazd, a city located in the hot and arid Iranian Plateau, where windcatchers have stood for centuries.[7] Like the CoolAnt, a bâdgir requires no electricity; it drives air, and the sand it carries, through the top of the tower and down into the structures below, with the sand deposited at the foot of the tower. The air is directed to flow over subterranean pools of water before journeying through the connected buildings, cooling their interior. As the air warms up, it is funnelled and released back through the top of the tower and the process continues. Variations of these towers can be found in Qatar, Egypt and Pakistan, but many now lay dilapidated. Once a sign of wealth and comfort, these traditional techniques of cooling were discarded for more 'modern' and 'fashionable' European designs and technologies.[8] There was a short boom in the construction of wind towers, with people returning to the craft during the oil crises of the 1970s and 1980s in cities like Doha; yet as energy prices stabilised, and electricity-generated cooling became affordable, investment trickled away from the traditional methods again. Though clay-based technologies are neglected in countries like Iran, designers like Siripurapu, technologists and Indigenous communities around the world are coming together to keep alive traditional innovations, building up the future of sustainable living straight from the Earth.

In the spring of 2021, I was honoured to join a group of incredible designers on the advisory panel of the Design Museum's 'Waste Age' Exhibition set to open later that year. The goal of the exhibition was to explore our relationship to everyday items, examining the role of design in creating a culture of waste and providing a platform to the design solutions that were combating the mistakes of the past and carving out imaginaries of the future. During our conversations, much attention was brought to the challenges of manifesting the values and ideals of the exhibition into its physicality – it would make little sense to create an exhibition through wasteful means. We thought a lot about how the materials that would make up the walls, tables and plinths could and should exemplify the sustainable and regenerative futures we were trying to platform. These conversations were not trivial and resulted in an incredibly considered exhibition design, brought to life by the talented team at Material Cultures, an organisation which brings together design, research and action towards achieving a post-carbon-built environment.[9] The team used three key strategies to build the exhibition: first, they reused elements from previous exhibitions at the museum; second, they constructed exhibition rooms using natural and biodegradable materials including wool, locally sourced clay and engineered timber; and finally, every wall or room was designed to be deconstructed and be fit for reuse elsewhere – one of these structures being an adobe brick wall. These strategies were not employed frivolously or for aesthetic pleasure alone. They were living experiments in how we might materially build a better world. Buildings and construction account for a whopping 37 per cent of global carbon emissions, with cement being a key culprit, its manufacture contributing 8 per cent of global emissions alone.[10] With many countries working to provide enough homes for their citizens to meet housing needs – in the UK the target is 300,000 by 2025 – the climate impact will only continue to rise.[11] The need for alternative materials that provide safe, efficient and comfortable housing without the carbon cost has become clear.

Waste Age had piqued my interest in sustainable building materials and practices. While the adobe wall in the exhibition reminded me of a very familiar image of traditional earthen structures, I knew very little about how it was being used to cut emissions in the construction sector. As I would go on to find, the wisdom of the past, applied in practices employed by communities around the world for millennia and even today yet largely ignored or devalued by the Global North, are the key to contemporary sustainability efforts.

Adobe is a natural building material often created from mixtures of earth, clay and straw and has been used across the Middle East, Western Asia, Africa, South America, Southwestern North America, and Southwestern and Eastern Europe over thousands of years. Sources suggest that fired clay brick is one of the 'longest lasting and strongest' building materials, thought to have been used since 5000 BCE.[12] They are still used in a multitude of African traditional building practices: from construction of the homes by the Batammariba ('those who model the earth') of Togo to those of the Gurunsi from Burkina Faso, which are waterproofed using the wastewater from shea butter production.[13] One of the most prominent modern African architects, Francis Diébédo Kéré from Burkina Faso, believes that clay 'connects the past, present and future'; when I learned how the traditional custom of clay construction was being incorporated into modern regenerative design practices through 3D printing, I knew exactly what he meant.

Between 2020 and 2021, the students of the postgraduate course in 3D Printing Architecture at the Institute for Advanced Architecture of Catalonia (IAAC) worked on a project called New Ghanzi. They investigated questions of social justice, heritage and sustainable design in order to provide housing for members of the San community in the Kalahari Desert in Botswana. The San, a nomadic community of hunter-gatherers, are the oldest inhabitants of Southern Africa and have lived in the Kalahari Desert for an estimated 20,000 years.[14] The name San, meaning 'foragers', came from the Khoikhoi nomadic cattle herders, a community that coexisted with the San, and the term was intended to be derogatory.[15] However, San became the chosen name of the community as opposed to the label of 'bushmen', which came from the Dutch colonial term 'bossiesman' meaning bandit or outlaw.[16] In fact, most members of the San community do not use either label and rather refer to their individual group names, which include the !Kung, Tuu, Tshu-Khwe, Khwe and others, and they live in the region that spans the countries of Southern Africa, including Namibia, Botswana and Angola. For consistency and clarity, I will continue to refer to them as the 'San' as is now largely accepted. The San are known for their incredible tracking skills, able to trail animals across pretty much any landscape and even able to tell if an animal is injured just from its tracks.[17] Since the colonial era, the San way of life, rooted in migration, foraging and hunting, has been destroyed, propelling them further into social exclusion, poverty and disenfranchisement. And while their traditions, skills and ways of life have attracted much attention from anthropologists and filmmakers, the idealistic and romanticised images of San life hide the realities of a community who are constantly disconnected from their land and practices and who have become some of Southern Africa's poorest people. Much of this disconnection has stemmed from the San community's lack of land rights, which has led to many enforced evictions, most notably from the Central Kalahari game reserve in 2002. After four years of campaigning, the eviction was finally judged 'unlawful and unconstitutional'; however, the disenfranchisement of the San people is widespread and multifaceted, continuing to this day.[18]

Recent developments, specific to the motivations of the New Ghanzi project, such as new anti-poaching laws and the discovery of diamonds in the Kalahari Desert in Botswana have led to further evictions of the San people, forcing them to find a new home. The students and faculty from IAAC set out to design and build sustainable, culturally appropriate and climate-responsive housing for the San community. The project had to respond to key climate challenges that affected the San community, including intense heat and large temperature fluctuations, especially in the winter months. One way in which the IAAC team approached this challenge was by creating microclimates by erecting courtyards between buildings that were inspired by the traditional San spatial practice of clustering circular dwellings around a central courtyard. Circular buildings are seen all over Africa and, with their reduced surface area, are very well suited to dry and hot climates. Unlike the surrounding building developments that erect singular dwellings separated from each other, the constructions under the New Ghanzi project embed the inherently social culture and infrastructure of the San in the connected dwelling design. To construct the houses, the team used the innovative WASP 3D printing technology, in which a large crane with a network of arms prints with clay and mud, accessed directly from the site, according to a preprogrammed design. The project was a collaborative one, and members of the San community were taught over a few months to use the crane to provide not only practical training but also tangible job opportunities for the future.

The New Ghanzi project and the CoolAnt terracotta air conditioner design brings us closer to the meaning of what rooted innovation really is – the guided development of modern technologies that seek to centre and serve the teachings and needs of nature and people. It is about technological solutions that are not imposed but connected, that do not destroy but support, that do not dominate but integrate. These projects beckon our gaze to the earth beneath our feet. To acknowledge the material nature of technology as something that comes from and carries with it the energy of the Earth and the sacrifices of its people, no matter how removed it may seem in its final form. To return to the Earth, to collaborate with the land, transforming the old into the new and making the new from the old. They show us how we can harness the power of modern technologies to work for, not against, people and the planet.

17 From Innovation to Theory

In his book *Small Is Beautiful: Economics As If People Mattered*, the late British economist E. F. Schumacher wrote that 'wisdom demands a new orientation of science and technology toward the organic, the gentle, the elegant and beautiful'. And to this I would add, more *rooted* approach to technology. Throughout INNOVATION, we have seen how we can place nature, the wisdom of the past and environmental justice at the centre of radical environmental innovations. We started by putting into practice the lessons we learned in IMAGINATION, challenging how we define technology and innovation, reorienting innovation as a practice of craft and care rather than extraction and domination. Whilst acknowledging technology's potential for destruction, and the role innovation has played in catalysing climate breakdown, we moved beyond simplistic, binary visions of technology as either saviour or villain. By rejecting the binary and engaging with stories that lie between past and future, traditional and technological, we find that innovation doesn't have to be at odds with the world and those that live in it. Reorienting ourselves and amplifying stories that bring us new visions of rooted innovations is to open ourselves up to think differently about how we build the solutions of the future, bringing squarely into focus the knowledge that nature itself is a technologist and that all technology ultimately comes from the Earth.

In this section, we have expanded and deepened our view of innovation, and we must bring forward these skills and apply them to our social and political worlds too. We are in desperate need of grounded cultural innovation.

New ways of reflecting on, thinking about and remaking the systems that govern our relationship with each other and the planet. In the next section, THEORY, we will come to see that by building strong coalitions and collaborations between scientists, designers, technologists, activists, mothers, children, lawyers and elders, we can change the social, political and economic systems our world is built on and the world itself.

Theory

18 Theory as Liberation

Theory is often resisted within mainstream environmental movements, discounted for its perceived lack of impact and seen as a distraction to the work of activism. Despite the fact that theorists – through research, science, law, economics or politics – influence much of our national and global climate action and policy, it can feel hard to understand what exactly they contribute to the environmental movement. There exists an understandable apathy or scepticism towards the individuals and institutions behind the numbers, facts, stats and frameworks that impact the decisions made about our future, and for good reason. Much of the work in these spaces is incredibly niche, usually communicated for the benefit of only those within or closely related to the respective field. In the Western intellectual tradition, we assume, as described by the scientist and writer Rachel Carson, that 'knowledge of science is the prerogative of only a small number of human beings, isolated and priest-like in their laboratories'. I am extending her use of science here to that of theory in general. That this work must be shut away and conducted by only those who have a 'right' to produce knowledge and influence society based on that knowledge. Oftentimes, the text is too impenetrable, the meaning lost in a sea of obscure words. This work is also seen as a perpetuation of systems of oppression and elitism – they serve and favour those who hold high socioeconomic and political power in this world. For those of us from marginalised backgrounds or from countries recovering from the impact of colonialism, the knowledge that many of the institutions that produce fundamental, world-changing theories are built on legacies of colonialism, discrimination, elitism and unjust power dynamics is also off-putting.

When the language of theory does enter activist or organising spaces, it is often used as a blunt tool to assert dominance and superiority. I have countless stories from friends, those who have never been involved in activism and those who are seasoned organisers, of attending meetings that were completely unintelligible. Forced to sit through hours of verbal jousting, often between men quoting passages from political theory books as if from a pulpit. Whether within the ivory tower or outside it, it is easy to resign ourselves to the thinking that no equitable or just change could be made with or through theory. But, as you will come to see, nothing could be further from the truth.

Before we go any further, let's take a few steps back. What exactly do I mean when I say theory? When you hear theory, you might think to the worlds of maths and physics, to theories such as general relativity or quantum theory. Abstract, far away and complex ideas brought to us by the inaccessibly intelligent 'founding fathers' of science. While this is probably the most accepted idea of theory, in this section, I speak of theory through an interpretation inspired by the etymology of the word itself and through the work of one of my favourite feminist theorists, bell hooks (more on her later). According to Oxford Languages, the word can be traced back to the Greek for contemplation and spectator. Theorising is simply the act of observing and then contemplating on what happens in the world. A simple act that holds huge power. There is power in the ability to visualise, describe and comprehend the world around us. Freedom in the ability to name the systems and structures around us, to understand our experiences within them and then be guided towards acts of resistance.

You can think about the act of theorising like discovering a heap of trash on a riverbed, and instead of just accepting that it should be there, wading steadfast through the water until coming upon the source of pollution. From tracing the effect on the ecosystem, and your health, to recognising the systems creating them. In Chapter 22, this simple example is brought to life by the team at The Or Foundation, who traced fashion waste and pollution from Kantamanto Market, one of West Africa's largest second-hand clothing markets, up the River Volta, one of the most important waterways in Ghana. Their work was essential in naming and shaming the Global North fast-fashion brands suffocating and polluting ecosystems in the Global South, and in mobilising local and global communities to take action.

For me, there are three key reasons why theory must play a significant role in our everyday lives and in our approach to environmental action:

1. What we don't know can hurt us: We often know the what. The what is the trash in the river. But theory tells us the *why* and the *how*. Without understanding these, we become either the victim of or complicit in the destruction.
2. Through theory, we can make better, more critical decisions: Once we know the why and the how of systems of destruction and our complicity with them, we can begin to make decisions and take action. We can, like the residents of Warren County, organise and resist the source of pollution. We can engage with the questions of animism that

arose in the section on IMAGINATION, and ask why our community does not see the river as a relative, leaving it suffocated rather than clearing it of the trash.

3. We must overcome the illusion that we don't have choice: Theory gives us the agency to make a change. Instead of just accepting that the river is polluted, and there's nothing we can do about it. Theory allows us not only to name the source of pollution and the pain it is causing, but the systems that enable the destruction to take place in the first place. Systems of oppression that are built on their own racist or discriminatory theories as we saw when learning about redlining. Theory is a tool to dismantle the myth that systems cannot be changed and to build the skills of comprehension and develop the craft to create alternative ones.

We don't *need* theory to explain our experiences back to us, but it does help us articulate what we know to be true in order to share that truth with others and to begin strategising on the ways we can enact change, together. I don't need theory to understand racism, I *lived* it. I don't need theory to understand sexism, I *lived* it. I don't need theory to understand, trauma, depression or anxiety, I *lived* it. Those who are experiencing the worst impacts of environmental breakdown, living through floods and forest loss, or who are desperately fighting for climate justice don't *need* to read about it, they are *living* it! When speaking with Miranda Lowe, Principal Curator at the Natural History Museum, she told me that finding roots in theory is about bringing to the fore 'the knowledge that [we] already hold'.

The role of theory is to resurface what we feel and experience, and translate it into what we can come to know, can communicate and can act on.

The person I turn to most often when thinking about the powerful role of theory in our lives and our movements is the late bell hooks. hooks is one of the most beloved and respected feminist theorists – best known for her work and teachings at the intersection of women's oppression, race, gender, sexual identity and class. Her 1981 text 'Ain't I A Woman' was foundational to the shift in perspective around the experiences of Black women within the feminist movement. Her work was grounded in reflecting the world through her theory and changing it through her actions. For hooks, this work was deeply personal and emotive, a notion unthinkable to those who hold more traditional Western – and limiting – ideas of what it means to be a scholar. The opening lines to her essay 'Theory as a Liberatory Practice' describe how she came to her work through pain. She said,

> *The pain within me was so intense*
> *that I could not go on living.*
> *I came to theory desperate,*
> *wanting to comprehend—to grasp*
> *what was happening around me.*
> *Most importantly, I wanted to make the*
> *hurt go away. I saw in theory then a*
> *location for healing.*[1]

hooks lived in a world that despised both her gender and her race. A world where the adults around her, exposed to myriad systems of oppression, had become disillusioned, apathetic and exhausted of their own agency.

Like hooks, many of us, whether because of race, gender identity, class or religion, face similar oppressive systems; the same systems that underpin and fuel environmental breakdown. These systems, as described by hooks, 'destroy our psychological and physical well-being' as well as the well-being of the planet. This destruction often leads us to act out of urgency, to act only for survival. Yet, from first-hand experience of living in a survivalist environment where critical thinking, questioning and curiosity were suppressed, hooks's attraction and dedication to theory was a rebellion. In her essay she explains that oppressive systems cannot be addressed by survival strategies alone, that we need 'new theories rooted in an attempt to understand both the nature of our contemporary predicament and the means by which we may collectively engage in resistance that would transform our current reality'. By theories, she is not talking about long Darwinian style treatises on the theory of evolution or Hawking's theory of black holes, as interesting as they are. No, theory here is the act of naming our pain, making sense of the world around us and thinking about what could be different.

In his influential book *Pedagogy of the Oppressed*, where pedagogy refers to the methods and practices of teaching and learning, the Brazilian educator and philosopher Paulo Freire tells us that, through theory, 'oppressed people can acquire a critical awareness of their own condition and . . . struggle for liberation'. Theory without action, he tells us, exists as only 'idle chatter' but action without theoretical reflection exists as 'action for action's sake and . . . makes dialogue impossible'. What he is describing here is the symbiotic relationship between theory (learning) and action (praxis). In this way, theory becomes an essential tool with which we can dismantle systems of oppression, which are also built on theories that reflect back the thoughts and beliefs of the powerful. Through theory, critical reflection, political education and collective wisdom, we can better understand the world and the ways in which we can and must transform it.

During my time running ClimateInColour, one of my favourite meetings to organise and conduct was our monthly ClimateInColour Reads, an accessible reading group facilitating interaction with foundational and ground-breaking climate and environmental research, essays, theories and frameworks. I started the group out of the desire to share the readings that were influencing my practices as a technologist in my PhD. What I found was a real hunger for accessible climate and environmental information. Each month, we would read academic articles, industry white papers and scholarly thought pieces, most of which members of the group had never accessed nor had the opportunity to engage with before. Over the years our group welcomed young people, neurodivergent people, academics and non-academics alike. We would spend each hourly session diving into our thoughts, arguments for and concerns with various schools of thought and think about how they might or might not be useful for a future sustainable world. From reading the original document of the US Green New Deal to an in-depth progress report on Indigenous Resistance Against Carbon and the Principles of Environmental Justice, we traversed continents, communities, themes and topics.

Each article, essay or report was accompanied with an audio file for those who wanted to listen to the papers, and I also produced in-depth, plain-word summaries so that regardless of people's capacity or time, they could take part in the conversation. These sessions were not just about consumption of information but, ultimately, about coming together, employing the lessons learned from our IMAGINATION practices, to ask questions about the research and our world. To propose new ways of looking at environmental and social issues and imagine new ways of doing things. Often, people would be able to link information from the papers to aspects in their own lives. By reading these works, lessons could be gleaned and reapplied to challenges people were seeing in their own workplace or their local communities. I felt privileged and encouraged watching people animatedly throw themselves into discussions on topics which excited them but which they would never have engaged with otherwise. They would have been held back by frustrating imposed and imagined barriers that prevent the layperson's entry into the seemingly 'elite' world of theory, thought to be reserved only for academics or those desperate to wield their 'intellect' like a sword, claiming superiority over others. The group was proof that theory can and should be accessed and constructed by a wide range of people and that theorists exist everywhere.

The truth is, so many of us 'might practice theorizing without ever knowing or possessing the term' and there are equally as many who practise theory but don't put their learnings into practice.[2] The work of a theorist – professional or organic – of analysing, synthesising, hypothesising and communicating, is an art. No, it is magic. It is to uncover, highlight, analyse and design guiding principles, frameworks and blueprints for life on Earth. Some of my favourite theories are proverbs. They are essential guides to life for many communities around the world, allowing us to hold and bring forward the wisdom and knowledge of our ancestors in to our present experience. When we take a more holistic view of what theory really is – turning away from images of lone scholars sitting by candlelight, hermits in libraries and institutions far away from the frontlines of activism and environmental action – we see that our experiences *are* theory. In an interview on why 'We Need Intellectuals' with Critical Resistance, a US organisation working to dismantle the prison–industrial complex, Angela Davis summed it up perfectly; she said, 'We need intellectuals, not just the professionally trained intellectuals but also, organic intellectuals' – those from all walks of life who think and reflect on destructive systems and craft new paths to lead their communities away from oppression. Our lives are theory itself. Davis tells us that 'if we are not in a position to reflect on what we're doing, then we will forever be stuck in one place'. Most of us are tired of this place, tired of loss, of exhaustion, of exploitation, of violence. The theory I speak of in this section involves carving out space for us to reflect on our experiences, to pause and engage more deeply with the systems utilised in the destruction of our environments and well-being and to learn from and employ the methods of those who have already encountered and resisted those systems.

Speaking to environmentalist and educator Isaias Hernandez (aka Queer Brown Vegan) on his own practices of theory as someone who has worked both within and outside of 'professional' spaces, he tells me that in life, we are constantly 'learning, unlearning and relearning' and that we must all 'stay curious and constantly ask questions' of the world and our communities. Engaging with this curiosity and cycle of learning, what we find is that theorists exist in universities, grocery stores, hospitals, schools, bars and prisons. The best theorists, in fact, may well be children, who bell hooks highlights 'have not yet been educated into accepting our routine social practices as natural' and so constantly question them, in a way often seen as impotent. Children show us the heart of theory, the dedication of constant curiosity and questioning. Of resisting acceptance of the world as it is *because someone said so*.

In this section, I want to bring to the fore the theorists who exist within and outside the 'professional' realm. Those who sit with their heads down and their hearts open, reflecting and scheming, through struggle, for our future. In this section, I want to uncover the historical and ongoing interweaving of theory, activism and social change. To create space for us to reflect on the interconnectedness of how we investigate the systems that cause and fuel environmental breakdown and how we resist them. Paraphrasing Nigerian author, philosopher and writer Báyò Akómoláfé, rooting in theory is about being led to new ways of being, outside of accepting 'business as usual'. This section will be a practice of theory, of documenting and contemplating the structures that perpetuate environmental breakdown and of exploring the foundational works that provide us more just alternatives. There exists a mighty community of theorists writing, connecting, speaking, teaching and engaging in cross-boundary, rule-breaking, paradigm-shifting work. Individuals and groups who are committed to connecting and centring their work in community as an act of resistance against the status quo. From de-growth to systems thinking, to collective liberation through litigation, this section is an exploration of the community of dedicated theorists and theories we can base our environmental action on to build a better world.

19 It Takes a Lawyer, an Activist and a Storyteller (and You) to Change the World

'It takes a lawyer, an activist and a storyteller to change the world'.[1] These are the defining words of the book *The Revolution Will Not Be Litigated: How Movements and Law Can Work Together to Win*, a collection of essays from '25 of the world's most accomplished movement lawyers and activists'.[2]

The quote reminds me of the story of sociologist Robert Bullard, fondly known as the father of environmental justice, lawyer Linda Bullard (also Robert's wife) and Northeast Community Action Group (a group of African American residents), who came together to oppose the dumping of 'sanitary' garbage in Texas. The year was 1979 and Robert had moved thirteen hours south from Iowa to Texas, where he had taken a position as assistant professor at Texas Southern University, a predominantly Black institution. Whilst Robert had planned to spend the next forty years of his life working as an urban sociologist and planner supporting Black communities in Houston, 1979 would mark the beginning of his four decade-long career in the environmental justice movement. That same year, Robert met his future wife, Linda, who was defending the Northeast Community Action Group, who were fighting against the creation of a toxic landfill site in their community. When Linda shared this work with Robert, his immediate reaction was 'You need a sociologist.' Already engaged in social issues within the Black community through his academic role, Robert teamed up with Linda, assuming the role of 'researcher as detective'. Their case, *Bean v Southwestern Waste Management*, was a first of its kind, not only in charging environmental discrimination (regarding the location of waste disposal as an extension of the Civil Rights Act) but in that everyone – Linda and Robert, the plaintiffs and expert witnesses – was African American. Despite the court acknowledging the obvious hazards and threats to health imposed by the waste disposal sites, it refused to accept that the location choice was based on racial discrimination due to lack of evidence of this. However, Bullard compiled masses of government records and conducted on-site visits and interviews, and reported that although the Black community only accounted for 25 per cent of the population in Houston between 1930 and 1978, 82 per cent of all the waste dumped in the city was in Black neighbourhoods. The research put into words the lived experience of the Black residents who 'had spent much of their lives escaping from waste sites, only to find waste-facility disputes following them to their new neighbourhoods'.

Bullard continued this research, motivated to understand how typical this pattern was. He extended his study area to four more Black communities, including those along the Mississippi River in Louisiana, the communities from which the name 'cancer alley' originated. What his research revealed was that the location of local landfill or waste sites was not random, but that 'in all cases, the residential character (e.g., the racial and economic profile) of the neighbourhoods had been established long before the industrial facilities invaded the areas'.[3] His findings also revealed that the largest hazardous-waste landfill in the US was located in Emelle, Alabama, where the population was 95 per cent Black. The conclusion? People in Black communities in the US are 'likely to suffer greater environmental and health risks than in the general population'. Motivated by the stark facts that stared back at him, Bullard compiled his case studies into what became a pioneering body of work in the environmental justice field, *Dumping Dixie: Race, Class and Environmental Quality*.

The book preceded a string of important action from Bullard and his peers, most notably the First National People of Color Environmental Leadership Summit in 1991. The summit was a 'watershed event', bringing together people of colour from across the US and facilitating the creation of the seventeen Principles of Environmental Justice and the Principles of Working Together. These two documents are important as they lay the foundations of environmental justice activism, influencing the work and approaches of many people involved in a variety of movements, including my own. Where the Principles of Environmental Justice map out the necessary scaffolding for a truly just world, the Principles of Working Together dive into how we actually create that world. Fundamentally, they present the idea that diversity in action is not just inevitable but essential in bolstering social movements, and that without understanding how we are called to act as a collective, what we do is weakened. We see this sentiment embedded in the third Principle of Working Together, which teaches that working together relies on our ability to recognise that movements are strengthened when there is participation and power distributed at every level. The *Bean v Southwestern Waste Management* case brought together community activists, lawyers and academics. No single role brought the case to life. It was conceived through the collaboration, motivation and action of many individuals, all differing at the source but coalescing in a collective goal for justice.

Over forty years later, the legacy of the essential work of Bullard, Linda and the Northeast Community Action Group in 1979 lives on. In 2023, a landmark environmental justice victory was won by Black residents of Lowndes County, Alabama, who the US Department of Justice said were first discriminated against, exposed to raw sewage and denied adequate access to sewage systems and then fined for the sanitation problems they were subjected to by the state.[4] The community have suffered the effects of racial discrimination and climate change. Catherine Coleman Flowers – the founding director of the Center for Rural Enterprise and Environmental Justice, one of the groups that filed the civil rights complaint against Alabama – highlights how 'climate change is bringing [racial discrimination] to the forefront, with rising water tables and intense storms that push more water through sewage systems', resulting in bubbling sewage in resident backyards.[5] The US Department of Justice launched the first ever Title VI environmental justice investigation, which was created as part of the landmark Civil Rights Act of 1964 and forbids discrimination based on race, colour and national origin in programmes and activities receiving federal financial assistance. The investigation found that the Alabama Department for Public Health was aware of these indignities, but failed to rectify the sanitation discrimination in the state.[6] In May 2023, justice was finally won for the Lowndes community when a settlement was reached. The Alabama Department of Public Health denied conducting its sewage or infectious diseases programmes in a discriminatory manner, but said it looked forward to implementing the settlement to benefit Lowndes County residents.[7]

We have seen how, in the US, the courtroom can become a place of collective protest, where citizens can advocate for their rights. Returning to our definition of theory as practice of collective enquiry and transformation, we see that legislation, when rooted in community, offers a powerful tool for securing environmental justice around the world. But the courtroom is also a conduit for the collective wisdom embedded in the advocacy for and protection of the rights of the non-human world too. In the next chapter, we will witness how the fusing of legislation, ancestral wisdom and community organising open us up to the understanding that theory is strengthened, not weakened, by feeling, emotion and love.

20 Theory as Collective Wisdom

It was late March of 2022, and I was on my first trip to Ghana. In the back of a rattly 4x4, riding the bumps and ditches of the rich clay road, I was full of anticipation. My eyes were fixated on the scene passing by my window, but my mind was on a journey of its own, flitting between questions, anxieties and ideas. My daydream was abruptly interrupted when the car suddenly started to slow down. I looked through the windscreen and saw that we had stopped in front of a tree in the middle of the road. My guide excitedly started to tell me the tale of this tree, which went a little like this.

In the midst of the Asante–Denkiyra Battle of Feyiase in 1701, Okomfo Anokye, the first traditional priest of the Asante Kingdom spat out the kola nuts he was chewing. He spat them out in frustration (or disgust) at the violence of his people. Just days later, at the exact place the chewed kola nuts landed, a tree began to sprout.

This tale is part of oral history and may vary depending on who is recounting it but the sentiment remains even if the story is open to shifting. Across Ghana, and other West African countries, the kola nut is sacred, used not only in ceremonies like weddings but also as a source of caffeine (one kola nut has more caffeine than two cups of Americano coffee), and is a powerful medicine for treating skin diseases. To this day, the tree in the tale stands, despite the ongoing development of a motorway around it. The essence of Sankofa is kept alive in that tree, which is kept standing through the power of cultural history and mythology when it could just as easily have been cut down, its protection founded in cultural history and mythology, despite human needs and wants. It is this cross-generational knowledge exchange through proverbs, traditions and myths that grounds the relationships many individuals and communities around the world have with the Earth. These proverbs are theories that each generation must study and then internalise so that it becomes part of their being. These theories become ways of life, inextricable from the experiences of students they imbue. You see, theories of culture, teachings of human–nature symbiosis, just like us, have a will to live on. Like seeds, they are dispersed in our minds so they may continue in this earthly cycle, growing into a mighty web of collective wisdom. It is there that they sit, poised, waiting and ready to be employed, if only we nurture them, when threat arises.

Ko au te Awa. *Ko te Awa ko au.*[1]
I am the river the river is me.

This short but powerful traditional saying expresses the Māori worldview that acknowledges the interwoven existence of water and humans. More generally, it is the manifestation of the Māori philosophy *whakapapa*, which can be loosely translated to 'connectedness'.[2] *Whakapapa* teaches that all Māori are born into a web of connection with human and non-human ancestors, including natural ecosystems.[3] For many Māori communities, rivers exist as arteries, delivering physical and spiritual sustenance for the people, wildlife and forests on their banks. This could not be truer than for the Whanganui Iwi people who live on the west coast of the North Island of Aotearoa, the Indigenous name for New Zealand. The Whanganui River has a strong presence on the island, making its way down from the mountains, twisting and turning through densely foliaged valleys of the Whanganui National Park, finally pouring into the Tasman Sea. It is the longest navigable river in all of Aotearoa and provides the country with 5 per cent of its electricity. The Whanganui Iwi people's relationship with the Whanganui River runs deep, underpinned by the concept of *Te Awa Tupua*, the profound meaning of which is hard to translate to English but can be understood as the inseparability of the people and all facets, spiritual and physical, of the river.[4]

Te Awa Tupua.
The physical and metaphysical river.[5]

However, the island endured a centuries-long programme of destruction and disconnection at the hands of British colonisers, with the Crown declaring ownership of the river. Driven by profit, the desire for ownership and the lure of the coal resting unexploited in the bed of the river, the Crown deconstructed the river into a patchwork of separate parts to be sold to the highest bidder. The stark contrast, and subsequent manifestation, of this Western worldview and system of knowledge was incredibly painful for the Whanganui, who despite the extractive policies of the colonisers never relinquished their right to protect the river. It was a result of this unending drive fuelled by knowledge not written but felt that the Whanganui achieved perhaps one of the most beautiful realisations of justice, attained at the intersection of

Western and Indigenous knowledge systems. What unfolded in Whanganui was a nearly two-century-long appeal for the granting of legal personhood to the Whanganui River, combining legal theory and Māori collective wisdom to bring about justice.[6] Legal personhood, sometimes referred to as environmental personhood, is a legal theory that grants non-human entities the same rights as humans. The theory gained prominence in the Western world in the 1970s when the legal scholar Christopher Stone wrote an article titled 'Should Trees Have Standing'.[7] Stone's article had a huge impact on the environmental movement in the US, starting conversations about rights of nature that reached the US Supreme Court. For many Indigenous communities, this formalised law is not only obvious but central to their interactions with the rest of the natural world. In the context of the Whanganui people, what they needed was a way to use this formalised theory, one that translated their collective wisdom to the Western world, to bring justice for their ancestor. They needed to convince the legal system that any abuse, threat or violence enacted on the river would be equivalent to violence against the Whanganui people themselves. Although nowhere close to an example of the deeply engrained collective wisdom of the Whanganui, fighting for the bestowal of legal personhood to the river was an essential strategy for the community to protect and reconnect with their beloved kin.[8] And in 2017 they did just that.

Instrumental to this win was Gerrard Albert, the Chair and Principal Advisor of Ngā Tāngata Tiaki o Whanganui, the formalised governance body representing the Whanganui Iwi people and continuing the ancestral work of protecting and managing the Whanganui River.[9] With a twenty-five-year career in environmental and resource planning specific to the Whanganui Iwi, Albert played a lead technical role in the negotiations that would result in the Te Awa Tupua Act, a legal framework that enshrined the personhood of the river in law. Albert was a graduate of his traditional Iwi university known as the 'whare wananga' or tribal houses of knowledge. He came to the law through his legacy, taking over from his uncle in the negotiations that had, at that time, spanned longer than a tribe member's lifetime. It was through this training and the teachings of his elders that he rooted his legal approach, one in which the river remained his first point of reference.[10]

Representing his tribe, Albert's work was driven by his personal experiences: growing up amidst the degradation of the river and his memories of playing in the discharge-filled waters as a child, scrambling up the banks that had deteriorated from shingle to knee-deep mud as a result of gravel extraction. It would be the traditions and worldviews he grew up with that would end up saving this treasured river. He had observed how Western formalised approaches to governance and law restricted the Māori from emoting, from expressing their spiritual connection to the natural world and the collective wisdom integral to protecting it.[11] Grounded and held by his tradition, Albert rejected the narrow-minded, rigid and oppressive frameworks being forced on his community. For him, his rooting in theory came from deep feeling; he explained that 'when you are dealing with a sick river, you feel sick yourself. Because that river is you and you are that river'. For Albert, this work was essential and was founded in the motivation to protect and heal his human and non-human kin.

For many theorists practising on behalf of the

communities they are connected to, the ability to embed vulnerability, connection and feeling into this work comes as a tonic. Those, like Robert and Albert, who come to this work from pain or oppression are led by their hearts to rewrite the histories of extraction enacted on their people to create more equitable climate futures. When research and advocacy are born out of passion, humanity and equity, they bring forth an exciting and beautiful world of possibility. Possibility for effective environmental protection in tandem with cultural prosperity. In this way, the practice of theory is seen as a transcendence of a purely academic–activist binary, beyond the professional or the organic. Theory intertwined with emotion, love, care and community challenges the status quo to enact radical change. Moving beyond these binaries we can, as described by Floor van der Hout, a PhD student in sociology and international development at Northumbria University, 'promote, acknowledge, and embody the emotional dimensions of doing research and activism [which are] key in bringing about radical social change'.[12] To some, these 'emotional dimensions' have no place in research, advocacy or decision-making, and I have had to confront the real discomfort that this emotion-led work creates. In the next chapter, we will witness how the activist/academic binary can be rejected to build flourishing collaborations for collective justice. Going beyond that binary allows us to acknowledge that many people working in academia, advocacy or policy are not made up of two halves but are an ensemble of feeling, facts, numbers and spirit. They are committed to combining the wonders of science, the levers of power and knowledge systems from all corners of the Earth to advocate for and bring to life the futures of justice we need.

21 Straddling Worlds for Resistance and Change

It was a wet, late October evening. The time of day when it seems as if a veil of tightly woven linen has been draped over the sky, casting a flat neither-light-nor-dark hue across the buildings below. Running a little late, I quickly parked my bike on the busy King's Parade in the centre of Cambridge. Dodging the clusters of disoriented and soggy tourists, I headed over rain-slick cobbles towards a colourful and lively crowd. They were gathered outside Senate House, the location where, if all goes well, I will eventually receive my doctorate. I joined the group just in time to raise my voice in unison with the rally chants:

> PEOPLE'S HOMES ARE BURNING TO ASH,
> SAY NO TO FOSSIL FUEL CASH.
>
> PEOPLE'S HOMES ARE BURNING TO ASH,
> SAY NO TO FOSSIL FUEL CASH.
>
> PEOPLE'S HOMES ARE BURNING TO ASH,
> SAY NO TO FOSSIL FUEL CASH.

In the group were people from a wide range of organisations, including Cambridge Climate Justice, Cambridge Student Action for Refugees and the Cambridge Stop the War Coalition. Yet, what tied most of us together, what had brought us out of our warm houses that day, was our role as academics, with many of us working as researchers in the climate or environmental space. The University of Cambridge is one of the top ten recipients of funding from fossil fuel companies in the UK, receiving an average of £3.3 million a year for the last six years from companies such as BP, Shell, Total, Equinor, Eni, Chevron and Exxon.[1] As our chants faded to silence, I was handed a megaphone; I made my way to the centre of the crowd, quickly bringing up the speech I had written earlier that morning to close the rally. Quoting Angela Davis who said, 'I am no longer accepting the things I cannot change; I am changing the things I cannot accept', I reflected on how the acceptance of the environmental and social devastation brought upon our planet by the fossil fuel industry and its insidious ties with research made the university 'complicit [in] . . . a system that prioritises profit over people'. We were gathered that day as just one of many actions to change that which we could not accept and to demand a severing of all ties between academia and the fossil fuel industry. We stood firmly in our positions, straddling our roles as both activists and academics, moving within and outside of our institutions to make, demand and seed change.

The author and stoic Ryan Holiday wrote in his book *Rules for Radicals* that 'a true radical must work within the system . . . if it is a system worth destroying, it is a powerful one . . . and the way to destroy a powerful enemy is to use its own strength against it'. His words stand in direct opposition to the popular teaching of Audre Lorde which maintains that 'the master's tools will never dismantle the master's house'. As an activist-academic – working with and through theory – my perspective lies in a grey zone, where conducting work within the system can be both fruitful *and* futile. As bell hooks wrote, 'Theory is not inherently healing, liberatory or revolutionary. It fulfils this function only when we ask it to.'[2] Whilst the overlap between academia and activism is rarely acknowledged outside academic institutions, the actions we took that October day were a small part of a long history of academic resistances led by a diverse community of activist-academics on campuses around the world. An activist-academic is an individual who promotes social change through academia and research, countering the traditional belief that science and research are inherently objective or apolitical. The emergence of the activist-academic's role coincided with the burgeoning student movement of the 1960s. College students in the US would protest many forms of injustice, from race to poverty. The action on university grounds culminated in a global mass movement of student protests in 1968 taking place across France, Mexico, Brazil, Italy, Japan, Northern Ireland and the US. These protests brought to the fore questions about the role of universities and academic institutions in the public sphere, with students using their voices to stand up for freedom of speech, poverty eradication and racial justice. The legacy continues to inspire action today, with thousands of students around the world setting encampments to resist the role universities are playing in the killing of innocent Palestinians in Gaza. The role of the activist-academic – embodied by Angela Davis, who continues to advocate for justice through her words and voice today, and Audre Lorde, whose writing has influenced generations of action – is one that sits at the intersection of research, communication, politics, law and social justice. The work we do often comes from a deep love for this planet and the people that live on it. For many of us, this work is more than just research, it's a way of using our skills to be part of a wider movement of individuals ushering in a more equitable and harmonious world.

In *Futures of Black Radicalism*, an anthology edited by Gaye Theresa Johnson and Alex Lubin, author and scholar George Lipsitz tells us that for the Black community, who have endured and resisted myriad oppressive systems and politics, their resistance can be seen as the manifestation of a 'deeply politicised love'.[3] This sentiment has also been expressed in the works of bell hooks who makes it clear that 'without an ethic of love shaping the direction of our political vision and our radical aspirations, we are often seduced, in one way or the other, into continued allegiance to systems of domination'.[4] It wasn't until reading through the anthology and better understanding the Black Radical Tradition (Tradition, for short) that I was able to see evidence of scholarship from my own community that was rooted in the love and rage I felt. The Tradition provides a framework for Black researchers and academics to stay rooted in the work of liberation through practice. Lessons that would inform my own view and practice of activist-scholarship. The Tradition was introduced by the radical theorist and activist Cedric J. Robinson as an intellectual movement based in the liberation of Black people. The Tradition developed as a form of resistance to the rise of racial capitalism in post-slavery America, which saw the fight for justice being co-opted by the attraction of rising up the social ladder within Western capitalist systems – the very same systems that continued to make life a misery for the most exploited in Black society. The practice of Black Radical Tradition manifested in the collaboration between Black intellectuals and deeply marginalised and oppressed

Black communities.[5] It was a resistance to the Western tradition of hoarding knowledge for only the rich or elite, and was a framework for getting to the root of injustice, of standing shoulder to shoulder with communities facing oppression and metaphorically of looting the institution of its tools used to perpetuate discrimination to strengthen social political literacy and action. The Tradition shows us how to embed justice and liberation in our research and action. Whilst Cedric Robinson is credited as the originator of the Tradition, it is the result of generations-old collective Black intelligence gathered from struggle and emerging from 'African culture, languages, beliefs and enslavement'.[6] As an academic, the biggest lesson I have learned from the Tradition lies in its deft approach to the grey zone of working 'within' the system. Black Radicalism teaches us that the power of radical scholarship lies in its ability to uncover the 'root causes and structural conditions' of injustices that impact vulnerable communities, and that the success of these enquiries relies on constant conversation and collaborations with the communities in question.

Whether it's feet on the ground and arms in the air or heads in books and eyes on screens, the activist-academic is an embodiment of the ways in which protests can be intellectual, and theory a resistance. Protest can become an intellectual practice of exposing systemic ills and transferring knowledge to the wider community, while theory can become a practice of dissidence and resistance. Activist-academics have an acute understanding of the knife-edge they occupy but are driven by the transformative possibilities that exist on that edge. They are attuned to the fact that, as hooks described, 'the academy is not paradise ... [but] learning is a place where paradise can be created'.[7] In some cases, before finding paradise, the academic-activist must work against the creation of hell on Earth.

In the US deep south, nestled between two of Louisiana's bustling metropolitan cities, New Orleans and Baton Rouge, lie 120 miles of land, cut right through the middle by the mighty *M   I*

S

S

I

S

S

I

P

P

I.

Just a one-hour drive from New Orleans lies the parish of St James, home to the seventy-one-year-old former teacher Sharon Lavigne. Lavigne is a descendant of civil rights activists and is the founder of RISE St James, a grassroots faith-based organisation fighting for environmental justice opposing the expansion of petrochemical factories in St James Parish, Louisiana. St James Parish is the only home Lavigne has ever known. She recalls memories as a small child, living with her parents off of the land, playing with cattle, pigs and chickens in their gardens and caring for them, and catching shrimp in the Mississippi with her grandfather.[8] Sadly, these joyful idyllic scenes are but a thing of the past for many families in the region. This region was formerly known as Plantation Country and used to be covered in fields of indigo, cotton, sugar ... and African slaves. Three hundred years later, the land and its people are still scarred by extraction and exploitation, only now it has a new name. Cancer Alley. Or, more recently, Death Alley. This strip of land has been transformed from Plantation Country to Petrochemical Corridor.[9] Today, more than 200 former sugarcane slave plantations are home to some of the most polluting petrochemical facilities along the river; it is a region where the majority of residents are Black and are forced to breathe some of the most toxic air in the country. In some areas, communities are living with cancer rates nearly fifty times that of the national average.[10] Beneath the choking skies and soil smothered in chemicals produced by these facilities, lie the ancestors of the communities they pollute. It has been documented that historically, between one and three cemeteries were created on each plantation by the enslaved to honour their fallen kin. Thousands of which have now been desecrated by the extractive industries that sit above them.

In 2015, in the St James Parish, two slave cemeteries were uncovered during a survey for a proposed expansion of a refinery owned by Shell. Three years later, a member of the petrochemical corporation Formosa Plastics, FG LA LLC (FG) announced that it would undertake construction on a 3.5 square mile microplastic production complex on land that they claimed they initially thought did not disturb any historical properties. However, in a report commissioned by RISE St James, compiled by the environmental consulting firm Coastal Environments Inc (CEI), four burials and eight potential grave sights were identified.[11] It is a legal requirement that petrochemical companies identify historical properties, including cemeteries, that would be threatened by their plans of development. The Formosa Plastics spokeswoman, Janile Parks, told the Associated Press that the company was following the law and had fenced off one burial ground they found on their property in the wake of the report. She also wrote that, 'FG will continue to be respectful of historical burial grounds and will continue to follow all applicable local, state and federal laws and regulations related to land use and cultural resources.' For generations, the Black descendant communities held fast to the knowledge of their ancestral sites against the tide of erasure but were constantly gaslit by the 'professional' surveyors and petrochemical company representatives, with much of their on-the-ground, investigative research being blocked by private landowners. But following in the Black Radical Tradition, they continued to fight. The collaboration between Forensic Architecture, a multidisciplinary research group based at Goldsmiths, University of London, and RISE St James brought together a community of multigenerational genealogists, scholars and local historians. Together they combined professional and organic theory, honouring their ancestral knowledge, to resist the desecration of those ancestors and the ongoing pollution of their families. Forensic Architecture were commissioned by the activist group and tasked to use their innovative research methods, including 3D modelling, pattern analysis and remote sensing, to gather evidence identifying the locations of their ancestor's cemeteries. Their research was thorough, creative and deeply revealing. They began their study by investigating the effects of the petrochemical industry on the health of people in the communities in the area, preparing a heat map showing the levels of pollution in the region. They followed this by taking ten years of data from a local weather station to model the spread of airborne pollutants from three dozen petrochemical sites along the Mississippi to show the 'scale and concentration of chemical gassing of communities throughout Death Alley'.[12] In 2024, these same communities were classified by the council of St James as 'industrial and ... future industrial', justifying the high pollution rates, invalidating the harm to the human life that called those 'sites' home. Forensic Architecture also uncovered that around 200 properties throughout the regions had been classified by the state as 'ready' for industrial development and available for auction, a fact that members of RISE St James were completely unaware of. Using maps from 1719 showing Indigenous territories prior to colonisation, Mississippi commission charts from 1894, aerial imagery spanning seven decades from 1940 to the present as well as interview transcripts from formerly enslaved people, Forensic Architecture created an interactive 3D rendering of several of the plantations that would have existed along the Mississippi.

For RISE St James, the research was a validating and cathartic process after years of the community being rendered invisible. Lavigne shared with the research group that she felt she was living 'on death row, just waiting to die'. It was with this fear, rage and despair that she and the team set forth to stop the petrochemical companies from digging up her ancestors to build more of the plants that were already killing her community. In 2019, the group successfully stopped the construction of Formosa's plastic plant, avoiding liquid hazardous waste worth £1 million from polluting the area every year.[13] For this work,

Lavigne, on behalf of the community, was awarded the 2021 Goldman Environmental Prize. Despite the win, their work to resist further petrochemical approvals in the region continues. Currently, detailed aerial maps, cartographic analysis, digital archaeology and simulations created by Forensic Architecture are being used in a court case against the parish council to fight for a moratorium on the construction of new petrochemical plants in the area.

The work of Forensic Architecture and the RISE St James community group shows us that rooting in theory for environmental justice is about connecting to a history of struggle to resist and eradicate oppressive systems. These are histories that belong not only to the Black community but all of us. As the Black Studies scholar George Lipsitz wrote, 'The Black in the Black Radical Tradition is a politics rather than a pigment.' It's a politics of freedom, justice and liberation – goals that we *all* must strive towards. In our final chapter of this section, we will dive deeper into how intertwined our economic and political systems are with oppression and how they connect the decisions we often unthinkingly make in the Global North with the suffering of those in the Global South. We will learn about the work of an incredible organisation that brings these connections to light and, as the team at Forensic Architecture did, cares for and centres community in and through research.

22 Systems Change, Not Climate Change

In the autumn of 2021, after an intensely beautiful yet frustrating week of marching, talking, interviewing, dancing, singing, praying and connecting at COP26, I left the gothic city of Glasgow – and the mighty River Clyde, 'the liquid ear' as the Scottish writer and essayist Kathleen Jamie referred to it in her poem 'What the Clyde Said, after COP26', that bore witness to the week's proceedings – and journeyed to the Scottish coast. After a short train ride, I arrived on the banks of the River Tay, once part of the great glacier that stretched across Scotland and extended into the North Sea.[1] Now, before spilling out into the ocean beyond, the River Tay passes an equally impressive monolith-like structure, the V&A Dundee. Boat-like, built on an epic scale, the museum holds works, exhibitions and community engagements that inspire creativity and change, and in that cold November week, it held hundreds of designers, researchers and economists for the Design Council's *Design for Planet* COP26 event. I was there, alongside many others, to build capacity within design research and practice to address the climate crisis. There were ideation sessions where we brainstormed, wrote, drew and discussed how to overcome extractive systems as well as talks from incredible thinkers and doers from around the world. One of the most memorable presentations during the two-day event came from the visionary economist Kate Raworth. It's not uncommon for the mention of economics to get reactions of people either rolling their eyes or stifling yawns, but when it comes to Raworth, it's near impossible to be anything but transfixed. With buckets of energy and passion, she took us on a journey of understanding the system of capitalism and the linear economy, and presented her model Doughnut Economics as an inclusive and holistic alternative. Calling in remotely, she began her talk by pulling from beneath her desk – like a magician – a garden hose pipe which she twisted, turned, bent and shaped to represent different economic systems. At first, it was an exponential curve, with the hosepipe starting out in the bottom left corner of the screen and extending rapidly back out of the screen past the top of her head, representing our current system of exponential growth. And then, the hosepipe was manipulated into an S-curve, a period of steady growth followed by stability and balance, the representation of nature's approach to growth. Growth, yes, but not perpetual. Nothing in nature, but cancers, grow unabated. With a simple hosepipe and a few enthusiastic words, Raworth had communicated to the audience the work and understanding of decades of green economists and system scientists: that exponential growth, the prerogative of a capitalist system, is incompatible with the finite system that is nature. Many of us know this. We have read the articles, think pieces and books highlighting the negative impacts of a capitalist system on people and planet. I remember picking up a copy of Raworth's seminal book, *Doughnut Economics*, from my mother-in-law's bookshelf, feeling inspired and energised to see a comprehensive model of an alternative economic system that encompassed the social *and* ecological. If you are unfamiliar with the theory, it essentially describes the world as a doughnut, with the hole in the middle representing social suffering, the hole of deprivation, as Raworth described. This is the space where people's needs are neglected, resulting in poverty, hunger and discrimination. In order to move people from this space, and raise their standards of living, economic growth is needed to push everyone past the *social foundation* and into the green zone, *the safe and just space for humanity*. But this growth, the way we provide safety and flourishing for society cannot proceed unabated. Here is where she draws the ecological ceiling, the upper bound of the safe space, that highlights the dependence of our social prosperity on the environment. Beyond the ecological ceiling lies disaster and it represents a collective *overshoot* of the planetary boundaries – the limits of the natural world to withstand pollution and resource extraction. It was exciting to read of such a system in the book, to be presented with a comprehensive and relatively simple concept that felt so actionable. Listening to her describe it in Dundee reawakened this excitement in me, but it also left me craving for more. At least within my own echo chamber, it seemed that most people were on board and invested with the idea of a new economic system.

Alongside Doughnut Economics are two other major economic theories: circular economy and degrowth theory. The idea behind the Circular Economy was popularised by Ellen MacArthur, the record-breaking yachtswoman who circumnavigated the globe in just seventy-one days, the fastest time in 2005. After retiring from the sport, and reflecting on her survival on limited supplies during her journey, she set up the Ellen MacArthur Foundation, 'a charity committed to creating a circular economy, which is designed to eliminate waste and pollution, circulate products and materials (at their highest value), and regenerate nature'.[2] Degrowth theory or the degrowth economy is an idea popularised by Jason Hickel, an economic anthropologist and author of the book *Less Is More: How Degrowth Will Save the World*, which Raworth described as a 'powerfully disruptive book for disrupted times'. Degrowth envisions a world in which major polluting industries and economies are rapidly contracted and gross domestic product (GDP) is rejected as a yardstick to measure human flourishing and economic development. I found these economic theories and campaigns not only compelling but inspirational. Many of us feel and know deeply that the linear, extractive economy we live in is flawed and no longer fit for purpose; yet, it is often hard to challenge a system that entraps us all. It was a last-minute lab visit during my first research trip to Ghana that opened my eyes to the work happening on the ground to resist deeply flawed systems of capital and consumption and bring about social and economic justice at the intersection of waste, fashion and community.

Too much clothing. Not enough justice.
Too often a consumer. So rarely a human.
It is time to recover. You are invited.

In 2022, whilst in Ghana, I endeavoured to visit as many local organisations as possible, wanting to connect with other climate conspirers on the ground in the motherland. One of my most memorable visits was to The Or Foundation's No More Fast Fashion Lab, the organisation from which the words that begin this section are quoted. Working between environmental justice, education and fashion development, The Or Foundation is on a mission to catalyse a justice-led circular economy by identifying and creating alternatives to the dominant wasteful model of fashion: 'alternatives that bring forth ecological prosperity, as opposed to destruction, and that inspire citizens to form a relationship with fashion that extends beyond their role as a consumer'.

The fashion industry accounts for a fifth of the 300 million tons of plastic produced globally each year and is responsible for 10 per cent of global carbon emissions – greater than the international aviation and shipping industries combined.[3] Globally, every year, between 80 and 100 billion items of clothing are produced, and of the fibres and materials drawn from the natural world to create those clothes, 87 per cent end up in landfill or incinerators.[4] The link between the fashion industry and capitalism is a stark one. As the team at The Or Foundation highlights in their Stop Waste Colonialism campaign:

> *Waste itself is a by-product of a culture of disposability and of an economic system that incentivizes a linear, or hierarchical, accumulation of value.*

Fast fashion companies are forever profit-hungry, looking to make ever higher profits year-on-year, their eyes set on the exponential graph of growth capitalism holds out like a carrot before them. As their production costs are squeezed and quality sacrificed, care for the environment goes out the window in order to generate this profit and make products more 'accessible' to the consumer, which in turn goes on to fuel more consumption. We are told to 'buy now, pay later', to accumulate more and more stuff and turn a blind eye to the economic, environmental and social costs we will eventually pay for later. Fast fashion defenders, or more accurately overconsumption proponents, argue that the low cost of clothing provides accessibility for those with low incomes, forgetting that it is the Black and Brown garment workers who are on the lowest incomes and who suffer under unliveable conditions to produce items that are destined for the trash. As Aja Barber makes clear in her book *Consumed*, 'People buying 5–10 garments of fast fashion a year rather than 50+ are not the problem.' Those that have the privilege and habit of purchasing clothes multiple times a month every month – and the influencers that encourage them to do so – invest in and perpetuate this pattern of overconsumption. Even those of us, who with good intentions donate our old clothes to local charity shops, contribute to the problem, with 70 per cent of second-hand clothing travelling from the UK to countries like Ghana.[5] Some of these items might make it to a large-scale textile recycling facility, but, in most cases, they end up in markets like Kantamanto, one of West Africa's largest second-hand clothing markets. The market has been lauded as the heart of Ghana's 'new age' fashion revolution, giving budding new designers access to an abundance of affordable materials, styles, colours and fabrics.[6] The market has catalysed an era of innovation for the creative minds of the city of Accra, Ghana's capital, yet the towers of clothes that dominate the market represent a global failure. While the market at first glance appears to be a treasure trove, a second chance for the neglected clothes of the world to find new owners, what we see in Kantamanto is only a fraction of the actual waste. Over 15 million garments arrive in Ghana every week, 40 per cent of which will never be fit for resale, resigned to live the rest of their lives polluting the local environment, or will end up fuelling landfill fires.[7] The clothes towers are the manifestation of greed, overconsumption and a culture of waste. These towers are, as is referred to in the local Akan language, *Obroni Wawu* – Dead White Man's Clothes. As explained by the research from the Dead White Man's Clothes research project, conducted by The Or Foundation's co-founders Liz Ricketts and Branson Skinner, the expression arose out of the remarks from local sellers in the market that only someone who had passed away could give up so many clothes – for the sellers, the concept of excess and waste at this scale was alien. But as we and the sellers are obviously aware, the clothes are not coming from the backs of dead foreigners but, in fact, from thousands of people around the world caught in the loop of purchasing more clothes than they could ever need and discarding them.

The work of The Or Foundation embodies all three components of a healthy and circular economy, those outlined by Raworth's model – economic, social and environmental. The No More Fast Fashion Lab is a hybrid community centre and circularity lab, located in the heart of Accra and neighbours Kantamanto Market. Entering the lab, my husband and I were met with a bright, airy, considered yet vibrant and open space. We were greeted by Chloe Asaam, the senior operations manager, and Sammy Oteng, the senior community engagement manager, who gave us a tour of the space. We passed a wall display of the clothing tags, featuring many popular high-street brands. Sammy explained that the display was part of a research project in which the team, working with Kantamanto Market sellers, audited and documented the biggest fashion waste producers. We were then shown the workspace provided for the *kayayei*, women from other parts of Ghana looking for a better life, who carry the large clothing bales, weighing 55 kilograms on average, on their heads from the port to the market. For every trip they make, they receive a dollar, and at the end of a gruelling day, they return to a concrete floor to sleep. The team made space for them as part of their remanufacturing hubs created to upskill the local sellers with seamstress training, allowing unsellable items to gain a new life and bring more money into the pockets of sellers. During our tour of the space, we had the chance to talk in more detail with the team about their Kayayei Research and Chiropractic Care project. In front of a display of X-rays, we learned of the extensive bodily injury that results from carrying such heavy bales, the damage inflicted on their spines. We learned of the team's collaboration with the Ghanaian chiropractor Dr Naa Asheley Dordor, providing 100 *kayayei* women with health screenings and examinations. Along the back wall was a bright and colourful display of some of these upcycled items, sleeping mats, cushions and furnishings, to be used by the local community. We were walked through the upcycling process, which included the use of an innovative machine built to shred old, second-hand, unsalvageable clothing so that they could be used as the fillings and cushioning for upcycled products. This innovative machine was designed and built by The Or team in collaboration with the Kantamanto Market sellers and metal workers from the nearby Agbogbloshie – the world's largest e-waste dump. When we throw away our phones, laptops and cables, they end up across the world in places like Agbogbloshie. The dump receives hundreds of thousands of tons of used electronics, mainly from Western Europe and the US, which thousands of Ghanaian and West African community members living in severe poverty sort, collect and try and sell. It is intense and exhausting work with residents, including children as young as six or seven, scouring the waste heaps for twelve hours a day to make enough money for their families despite the extreme risk of respiratory illness that comes from high exposure to e-waste. But from this shredding machine – a machine built from the rubble of destruction, by the hands of the most affected, in a space centred on care – emerges items for rest and comfort.

Astoundingly, this is not where the work of The Or Foundation stops. True systems change requires holistic intervention across the economic, social *and* environmental. Boarding *The Woman Who Does Not Fear*, a 36-foot solar-powered, custom-built, aluminium catamaran, made by Benlex Marine Systems, The Or Foundation board member Yvette Tetteh, along with a crew of documentarians and scientists, embarked on a month-long scientific expedition along the Volta River in Eastern Ghana. The Volta River is the principal water source for 24 million people; the expedition was motivated by research conducted in Accra that uncovered the impact of textile pollution on water quality and the need to highlight the impact of overconsumption and fashion waste on essential waterways like the Volta.[8] Bringing together research, citizen science and storytelling, the expedition saw Yvette swim 450 kilometres of the river – around six to eight hours every day – collecting water samples and engaging with affected communities along the way. For Yvette, her motivation went beyond scientific interest and was embedded in the cultural drivers of policy change that could bring liberation and safety for communities in Ghana. She told me, 'In order to have cultural change, we need people to care, and I felt that my body could be this vehicle of care.' For Ghanaians, in any endeavour relationship-building and commitment is key, and without showing a level of willingness to be deeply rooted within the communities and ecosystems being researched, science and theory remain an elite practice. *The Woman Who Does Not Fear* was a way of bringing the lab out of the formal institution and right into the centre of cultures and ecologies being studied. It was also a way to ensure that the science, as Yvette made clear, 'did not serve artifice but was grounded in something tangible', engaging with the river, the pollution and the people in an intentional and embedded manner. This approach is not only for the benefit of the affected communities but also for the integrity and impact of the research. Reflecting on the intersection between data and lived experience, Sammy described how their research is just a reflection of 'the lives in the realities of these communities . . . they have lived the data that we are collecting'.

The work of The Or Foundation is a brilliant example of how we turn theory and science into action, using them as scaffolding to bring into being vastly different systems. Where some see only profit and greed or degradation and despair, those like The Or team conceive schemes of recovery, repair and restoration.

23 From Theory to Healing

Our explorations of THEORY, the role it plays in our lives and our movements, have taken us far and wide. We have visited the traditional courtrooms and ancestral rivers of the Māori, attended the activist-academic rallies of Cambridge and journeyed along the Mississippi to witness how researchers and grassroots community groups resist petrochemical oppression. While the stories of this section present a wide breadth of lenses through which to see theory, they are all rooted in the understanding that an essential and often invisible part of creating change is enquiry and reflection. The ability to name the pain that we humans and the non-human world are experiencing, to follow the pain to the source, illuminating it and building strategies of resisting and transforming it. What space do you give, within your mind, your community, your organisation, to reflect and scheme? To ask difficult questions and follow them all the way to their conclusion, no matter how scary or unsettling they may be?

When we feel or fear pain, it seems like the most comfortable thing to do is hide, to run away. We are witnessing the world crack and break in myriad ways, and our instinct is to plaster over the wounds, desperately hoping they will heal out of sight. In the next section, HEALING, we will come to see that overcoming crisis requires us to collectively move through the cracks. To see the wound not as an abyss, but as an opening that brings us closer to harmony with each other and the Earth.

Healing

Healing is about stripping away, making ourselves naked and entering back into the ecosystem.

—Obatala Efunwale

24 The Earth Is a Church: Ethiopia's Architecture-Inspired Conservation

Deep into the 2020 lockdown, as the days wore on and my boredom grew, I found myself turning to the online world and repository of *Emergence Magazine*, an online and print publication that connects 'the threads between ecology, culture, and spirituality'. One day, after hours of reading, I came across a mixed media – visual and written – story that captivated me. At the centre of the screen lay a circular, brightly painted roof, patterned with concentric circles of red, yellow and green – the colours of the Ethiopian flag. As the high-definition drone footage zoomed out, the circular roof was contextualised, surrounded by a small, stone-walled circular courtyard, and then, an expanse of green, a forest. Deep, neon and warm-toned greens, every now and then interrupted by the stark white of bare branches. Zooming out further still, past the forest and the second stone wall that enclosed it, a sea of parched brown agricultural land. What I was viewing was just one of the more than 30,000 Church Forests that adorn the mountains of Ethiopia. Captured beautifully by the cinematographer Jeremy Seifert and writer Fred Bahnson for *Emergence Magazine*, 'The Church Forests of Ethiopia' was just one of many reports, articles, studies and documentations of the incredible conservation phenomena created and sustained by the Ethiopian Orthodox Tewahedo Church – one of the oldest in the world. For brevity, I will go on to refer to the Ethiopian Orthodox Tewahedo Church as The Church in this section. Whilst preceded by paganism and Judaism, religions thought to have been adopted through commercial connections with the Middle East, Christianity in Ethiopia dates back to the fourth century, first introduced in the royal court of the Aksumites by captive merchant sons from Tyre, Lebanon, and spread throughout the country by Indigenous scholars.[1] Forest conservation and the sanctity of the natural world are embedded in the Ethiopian Orthodox Tewahedo faith, motivated by strong taboos and social sanctions for forest degradation.[2] Unlike Western denominations of Christianity, The Church perceives nature in a holistic manner.[3] Their understanding of nature includes not only the trees, plants and forest or the animals and waterbodies but also humans and the Ethiopian nation itself.

Between 1994 and 2001, Ethiopia saw its forest cover drop from 40 per cent to a minuscule 4.2 per cent due to agricultural expansion and overgrazing.[4] Much of the remaining forest was found encircling the thousands of churches, where the sanctity of nature is not only believed but manifested into action through care and conservation. Yet, the forests were still under threat. Although those within The Church embodied their spiritual beliefs in the way they tended to the land, they had very little control over the actions of those beyond their holy walls. The churches themselves were protected by stone walls, but the forest had no defined borders, exposed to and unprotected from the need of surrounding farmers. Little did The Church know that over 5,000 miles way, this same question, of how to protect and expand Ethiopia's forests, was plaguing the mind of a young researcher, Dr Alemayehu Wassie Eshete, who found that in order to study and regenerate the forests of Ethiopia, he would have to return home and go to church.

Dr Eshete was born in South Gondar, the former capital of Ethiopia, bordered in the north-west by the famous Lake Tana. He is a respected academic who studied both in Ethiopia and Sweden, and his work on The Church Forests has been essential for their conservation at the community and state levels. After completing his master's degree on The Church Forests while travelling between Sweden and Ethiopia, he went on to put them at the centre of his doctoral research. Reading through his thesis, I was warmed by the balance of his academic rigour and heartfelt gratitude for The Church: He dedicated his PhD to The Church's 'scholars for generations of dedication and faithfulness to the ... surrounding forests, faithfulness without which, the forests would have disappeared'. His ecological studies found that The Church Forests hold an abundance of biodiversity, with one home to over forty-six species of tree alone. In his writing, Fred Bahnson recalls an elder describing how 'a church without a forest is like a naked person. A disgraced person'. With the belief that The Church must be 'enveloped by a forest' – their very own Garden of Eden, entrusted to them by God – The Church Forests were seen as an ark, offering a last home for the plants, wildlife and humans escaping harsh conditions in order to find solace.[5]

Identifying these 30,000 Church Forests, each measuring no more than 10 hectares, Eshete began to envision a bold and ambitious solution to address the country's dwindling forest cover.[6] If he could bring together and connect The Church leaders, might there also be a way to connect the constellation of forests scattered across Ethiopia's mountains? With the support of his American collaborator Dr Margaret Lowman, Dr Eshete did just that.[7]

He organised an annual workshop bringing together the priests from various Church Forests and presented to them satellite images that depicted the ongoing forest degradation occurring around the places of worship. Through these sessions, a beautifully simple solution emerged from the priests themselves, inspired by the spiritually informed architecture of The Church.

The structure of the Orthodox churches is composed of three concentric rings. The innermost is the Holy of Holies, accessible only to priests and the resting place of the sacred *ark* or *tabot*.[8] The next ring, the Keddist, is reserved for the most dedicated and faithful, those who take part in regular fasts and take communion. The outer ring is the Qene Mahelet, used for the general congregation. Further than that, stands the church wall, within which people can also worship. The forest, enveloping these concentric rings, created a ring of its own, a safeguard of sanctity. The centre of The Church at the centre of the forest. The priests saw an opportunity to solidify the circular threshold of the forest and proposed that a stone wall, identical to the one surrounding The Church, should be erected around the entire forest. This technique was applied to fourteen Church Forests, with some regenerating so well that the walls had to be re-erected, giving the forest more space to breathe and grow into. In this way, with the dynamic ebb and flow of enclosure, protection and expansion, Eshete found a way to reinstate Ethiopia's forests.

Though a scientist, for Eshete, this work was more than just a professional or even ecological endeavour – it was his 'emotional and spiritual connection'. The Church forests and the work of Eshete teach us of not only the beauty but power and impact of holding and retaining sacred spaces.

Of being led by a collective understanding of the spiritual richness of the natural world that offers far more culturally than any capitalist imaginings of the natural world could conceive. We learn how collective spaces of worship and healing have the potential, when rooted in respect and reverence for the planet, to catalyse essential environmental work, bringing together hands in prayer and protection. We see how, in the search for spiritual practice, we can be led and reconnected with the natural world, engaging in a symbiotic process of healing.

The Church Forests also teach us about the tenacity that collective faith brings. Even though the Church Forest pockets hold vital biodiversity and many of them have evaded disastrous degradation from ongoing agricultural expansion, they currently exist only as small emerald jewels embedded in a barren and scorched landscape. For those of us without a practice of faith, this fact might be discouraging. How many times have I heard someone say, 'There's no point recycling your plastic or using a reusable cup, it's inconsequential when you think of all the pollution happening in the world' or words to the same effect? It often feels like those who have not taken much action at all are eager to condemn others' faith and environmental motivations as useless and futile. To declare the fight lost before it has even begun. But not the leaders and worshippers of The Ethiopian Orthodox Tewahedo Church. What their unfaltering faith teaches us is that when our action is rooted in something greater than ourselves, rooted in deep faith and purpose instilled through spiritual practice, regardless of religious denomination, our acts cease to be tied to an immediate result, where the threat of failure is seen as disaster. When we centre our environmentalism in the mutual healing of ourselves and the planet, our actions become an embodiment of our love and respect for the living world that houses, nurtures and nourishes us. As environmental regeneration activist and folklorist Daze Aghaji shared with me, faith teaches us 'to be able to sit with loss, to not become destabilised by it, but rather curious about the work that is needed to become unlost'. She reflects on the way that 'the land holds grief but still it persists', and so we must too.

We have lost connection to practices that allow space for collective grieving, collective mission-building and collective healing. Practices that go beyond religion and spirituality as doctrines of pain and rigidity or beyond new age hedonism and fancy but that are the acknowledgement of the sacred in all beings on this planet. As Emmanuel Vaughan-Lee, filmmaker, editor of *Emergence Magazine* and Naqshbandi Sufi teacher, shared with me, we must 'redefine spirituality outside of conventional religious texts that "other" and remove people from spaces of true connection . . . and focus more on the universal values present in all forms of worship and meaning seeking'. It's about moving away from religion and spirituality in the context of power, hierarchy and patriarchy, which are antithetical to healing, and instead holding space for the reckoning that 'our crisis of disconnection is disconnection from the sacred'. It is to reclaim the collective understanding that we, every one of us, is nature and to see the sacredness of the living world in all that is around us. In her book *Sacred Nature*, author and former Roman Catholic nun Karen Armstrong argues that it is within spirituality and religion that we can find the strength to process our grief and begin a journey of healing. She highlights the importance of taking a holistic view of the climate crisis, not only focusing on the 'carbon crisis', but understanding climate breakdown as the result of connections between culture, belief systems and the myths we as a society have told ourselves. To acknowledge that 'while it is essential to cut carbon emissions and heed the warnings of scientists, we need to learn not only how to act differently but also how to think differently about the natural world'.

We often speak of nature as a passive object. An object with the sole purpose of providing us pleasure. We are told that nature happens elsewhere. Somewhere far away. It is only when we travel for miles that those sparks of awe return. The goosebumps on our arms, the disbelief in our eyes that something so beautiful exists on the same planet as us. The heart-swelling, tear-pricking gloriousness of it all. Of Nature. Emmanuel makes clear that to see the sacredness in all things, we must go beyond a romanticised notion of 'the wild' and 'see the sacred in the world writ large, in concretized forms as well as forested landscapes, in the urban as well as in the country'. As he puts it, we must 'expand the gaze of where we think the sacred lies or we risk disrespecting the essence of the sacred', a force that exists not only in the ancient forests but also in the pollution-obscured New York night sky, the rain drops that slide joyfully down our faces and the silent songs of a park at midnight. Reverence for the natural world must become something innate and not only that which is awakened when we are removed from our daily contexts. When speaking to me, the traditional Nigerian spiritual practitioner and teacher Obatala Efunwale tells me that the sky is the ceiling to his church, and I am inclined to agree. The

Earth is a church and we must see its sanctity in the folds and wrinkles that adorn our bodies and also in the trunks of trees. In the calm then tumult of our feelings and also in local streams and riverways. In the cool long breath we take to steady our mind and also in the assuring exhale of the wind. In the squealing of children and the chorus of songbirds. In the most miniscule and the most magnificent. We must come to acknowledge and honour these truths, as invoked by George Washington Carver, by 'tuning in'.

This section seeks to explore how we might reimagine our connection and relationship to the planet we call home. To explore the act and art of healing, how we can access it; it also seeks to amplify and honour those that facilitate it. This section, at its core, seeks to engage with and explore the work of those with the steady, calm voice of inspiration that draws people to find peace and fills them with whispers of the sacred living world. In the thickets of trees, halls of mosques, naves of churches and on the banks of rivers, healing work is being done. As Karen Armstrong makes clear at the end of the introduction of her book, 'It is not a question of believing religious doctrines; it is about incorporating into our lives insights and practices that will not only help us to meet today's serious challenges but change our hearts and minds.' By reflecting on the practices of faith, eco-spirituality and communal grieving that allow us to reconnect with the living world through deep reverence. To touch the Earth as the Earth touches us and build a world that is healthy for us *all*.

25 Nature Is a Human Right

I have come to the log bench at the end of the riverside path in Paradise.
My feet squelching in deep ridges of mud, rivulets of water washing a thin coat of black and brown on the toes of my wellies.

On this journey, all 5 minutes of it, I not only walked, but bathed. Bathed in the songs of the birds. A cacophony of trills, chirps, squawks and whistles.

I sit on a sturdy log and something rustles behind me. Perhaps one of the muntjac deer that frequent this place. More likely, a pigeon preparing to take flight.

I hear blackbirds, siskins, house martins, chiffchaffs, great tits, wrens, and robins.

A sharp March breeze rushes through bare but almost-budding branches, disrupting the cascade of tree-top melodies. Then the air is still once more, and I bathe myself again.

To my left, I am kept company by one of the many willow trees that call this place home. But this one is special.
For no other reason than I see her most often. Three springs I have walked by her, twirled within her and felt her long limbs christening my head with love and tenderness.

A robin perches, seemingly out of nowhere, settling on a nearby branch. We make eye contact for just a second, and then he is off.
The river is high and murky. Not bursting like it was last week, but the heavy rains have left their mark. Specks of light dance just on the water's surface, travelling with the current, northwards into the city.

My brain was all a-muddle when I entered this place. And I can't say I am now sufficiently un-muddled – and this is okay.

But something feels lighter, softer, slower. I have been touched, and the aftermath will last, maybe for the rest of the day. Maybe until tomorrow.
Until I return.
Return to be bathed.

I wrote the above diary entry in late March 2023 on my morning walk in the Paradise Local Nature Reserve in Cambridge. In just two weeks, I would be back in Ghana, conducting the second half of my field research in the forest. Blissfully unaware of the destruction I would soon come to meet and too distracted by my own struggles. I was not having the greatest start to the year and the stress and anxiety had caught up with me. In what seemed like a cruel mirroring of the previous year, where I went into surgery just a month before travelling to Ghana, again I was ill. This time, thankfully, only a kidney infection. My depression and anxiety oftentimes present as agoraphobia – an extreme or irrational fear of entering open or crowded places, of being in places from which escape is difficult. Or of leaving one's own home. For me it is the latter that I regularly experience, which, as someone who has a love and need to be connected to local natural spaces, is incredibly paradoxical and frustrating. Ironically, despite working in a field focused on the protection of the natural world, I all too easily lose my deep emotion and sense of awe towards it. Overwhelmed by expectations, work, deadlines, commitments, anxieties and traumas. When my mind becomes engulfed, I often forget what brings me the most peace. All too often when we're walking through a local park, a local nature reserve, a community woodland or coastal cliff, our mind is abuzz with a myriad of thoughts. *Did I turn the oven off? Oh crap, I forgot to send that email! I must get back soon, so much to do.* Still more often our ears are shut off by headphones blasting our regular podcast or favourite song. But what of the birds? What about their song? What poems, jaunts and messages do we miss when we focus on passing through rather than *connecting to* the natural world around us? How often do we make time for intentionally slow, mindful walks, more crouching in the mud to observe insects? Standing. Dead. Still. In the middle of the woods, letting the songs of the birds wash over us? Sometimes, as I work, I open the soundscape files I recorded with community members in Ghana. As twelve-hour-long recordings of the Ghanaian forest play in the background, gradually I notice my heartbeat is ever so slightly slower, my breathing less shallow. It isn't until seemingly simple moments – leaving the house, putting one foot in front of the other, armed only with a journal and a pen, and entering a space of green – that I remember. It isn't until then that I remember what heals my heart.

In 2021, I was part of a judging panel for the Wellcome Photography Prize. In the mental health category shortlist was a powerful image from the artist Dulcie Wagstaff who photographed herself in her garden, head buried in the soil. For her, the image was a representation of her coping with her depression, of 'connecting with the earth whilst hiding from the world above'.[1] This imagery and Wagstaff's practice are not abstract but represent proven approaches to healing through the natural world, and specifically through the soil. In 2007, researchers from the University of Bristol and University College London found that interacting with soil microbes leads to increased serotonin levels.[2] And in the Covid-19 lockdown, researchers from the University of Essex studied the impact of community gardening on a group of people with varying mental health issues, finding that 'sowing seeds, weeding and tending plants and flowers were beneficial to improving mental health, with the group self-reporting that their satisfaction and wellbeing levels had increased by 9 per cent'.[3] These findings confirmed what many gardeners and children know to be true, that playing in the dirt is healing work.

There exist countless practices, anecdotes and studies alike that bring to light the link between the improvement of our mental health and connection to the natural world. I remember first coming to this understanding through the practice of grounding or earthing, connecting to the Earth's electrical energy by walking barefoot or lying in the grass for extended periods of time. Although understudied, grounding can, as has been found by some researchers, lead to a boosted immune system, reduced inflammation and improved mood and mental health. In her wonderful book *Losing Eden*, Lucy F. Jones highlights the link between stress, emotional regulation and the environment, explaining how the sympathetic nervous system, the main function of which is to stimulate the body's reaction to stress, is positively affected by the natural world, its smells, sounds and visual beauty. Surprisingly, she quotes research that shows that even people under anaesthesia have been found to produce fewer chemical biomarkers associated with stress – such as amylase in saliva – when made to listen to a recording of soft wind or birdsong.[4] These and other studies have ushered in a new era of therapeutic practices, such as Forest Therapy, that sees 'health as the result of an adaptive process of the human being to his physical and social environment'.[5] Forest Therapy has been found to support patients experiencing depression, anxiety and PTSD far more than hospital- or urban-based therapeutic exercises and was largely inspired by the Japanese practice of Shinrin-yoku.[6] A term coined by the director of the Japanese Forest Agency, Tomohide Akiyama, Shinrin-yoku roughly translates to 'forest bathing', a practice involving bathing 'in the environment of the forest, using all your senses to experience nature up close'.[7] At first introduced to get the public of Japan to be more active, several studies proved the psychological and physiological benefits of the practice, and it soon became accepted as a form of preventative care.[8] While Shinrin-yoku has been efficiently embedded in Western medical practice and enshrined in formal research, modern approaches to forest bathing have become removed from its cultural roots within the Japanese Shinto tradition. Emerging from Buddhism, Confucian philosophy and Japanese shrine worship, the Shinto tradition holds nature as sacred, with many forests also considered shrines, a tradition common in cultures around the world, such as in Ghana where they are referred to as sacred groves.[9] The proliferation of practices emerging from the Shinto religion demonstrates how 'understandings of nature and the environment – whether "religious" or not – are not merely abstract ideas: they influence and are influenced by daily life practices, social relations, and ways of using space'.[10] There are many other related yet distinct ways of knowing, such as Chinese Traditional Medicine or the Indigenous cosmologies of North America, Africa, Asia and the Pacific, that are rooted in the understanding that the health and healing of humans is interwoven with the health and healing of the Earth.

For activist and author Tori Tsui, 'Nature has always been a salve amid the chaos of a troubled world.'[11] In her book *It's Not Just You*, she addresses the oft-felt and cited experience of eco-anxiety and its intersectional causes, and also how we might move beyond its limits as a term that does not account for the ways in which environmental breakdown is 'hinged on so many pre-existing inequalities'. This question of how we understand and address disconnection from the natural world amidst climate breakdown is essential, and this section would be incomplete without a deeper exploration of the issues such as lack of access, discrimination and racism that hinder a truly collective healing experience. When we speak of connecting to the healing power of nature, we must be precise with who *we* is. With 50 per cent of the British countryside owned by 1 per cent of the population, how can we have conversations about communing with the natural world when it is inaccessible to so many? How do we begin to make the shift towards reconnecting with the natural world when it often seems that there are those who are more deserving, given their social status, background and knowledge, of connecting with the natural world? More deserving of receiving nature's love and witnessing nature's wonders.

I am lucky enough to buckle down for a long writing session in my bedroom, window flung wide open, and can hear the constant breeze between the leaves of the dozens of trees in my neighbourhood. I am lucky enough to hear birdsong when I wake and to be minutes away from a beautiful, albeit small, local woodland. I was even lucky enough that the council flat I grew up in, in north-west London, backed on to and overlooked a brook, and nearby I could access open fields and the Grand Union Canal in Uxbridge. Yet, there are many who live in cities or neighbourhoods with such loose planning and development policies that every last square inch of land has been, is being or will be built on. I have read the stark statistics of how nearly 50 per cent of neighbourhoods in England are condemned to having tree cover of less than 10 per cent.[12] This is not just an issue of aesthetics. As I write, Rome is experiencing temperatures of 40°C and receiving water aid from the state, temperatures in California have already reached 48°C and 52.5°C in Sanbao in Xinjiang province in China.[13] Recent research has shown that trees can cool cities by up to 12°C in the summer, and given we are firmly in the era of extreme heat, low tree cover can only be seen as a severe public health issue.[14] We cannot talk about healing our individual and collective relationships with the natural world when so many continue to be deprived of environmental security.

Those of us who do not live in the countryside or

close to natural landscapes, yet are lucky enough to have financial and logistical privilege, are able to momentarily suspend our disconnection by travelling *to* nature, to the 'great outdoors'. Something that became an essential lifeline for many during the Covid-19 pandemic. Many urban dwellers, including me, who had family to run away to in the countryside, fled into the arms of the natural landscapes they knew and loved. During this time, there were many debates about diversity in natural spaces in the British landscape and diversity of people visiting, working in and enjoying natural spaces that arose from the stark differences in access to outdoor solace during the depressing lockdowns. In 2021, the 'Out of Bounds: Equity in Access to Urban Nature' report, published by the community charity Groundwork, found that Covid-19 widened the inequality gap, hitting people from low-income areas, people from ethnic minority backgrounds and disabled people the hardest.[15] The report found that 40 per cent of people from ethnic minority backgrounds live in the most green space–deprived areas and 29 per cent of people living with a long-term illness or disability had not visited a natural space in the previous month. Across the pond, researchers were discovering the links between nature-depleted neighbourhoods and higher instances of Covid-19.[16] The research team quantified inequity in greenness and the proximity of communities to parks across all urbanised areas in the US and linked the results to Covid-19 cases across zip codes in seventeen states. They found that 'areas with majority persons of colour had both higher case rates and less greenness'. They took their research further and controlled for socio-demographic variables. What they found was that an increase of 0.1 in the Normalized Difference Vegetation Index (a metric that approximates the 'greenness' of an area of land imaged using satellites) was associated with a 4.1 per cent decrease in Covid-19 incidence rates. In short, small increases in greenness were associated with significant reduction in the contraction of Covid-19. As the world marvelled at how nature was 'bouncing back' in response to global lockdowns, and the privileged were able to escape into the solace of their loved landscapes, hundreds of thousands remained confined to unsuitable, inadequate environments, with no 'nature' to be held by.[17]

It was during this time that I was commissioned to write an article for Grizedale Forest, an area of woodland in the Lake District in England. The Lake District is a magnet for nature lovers, offering incredible hill walking, expansive views and a deep sense of connection to 'the great outdoors'. A few years back I had the opportunity to visit for the first time with family and it was glorious. Yet, beyond the influx of tourists kitted out in Patagonia, the North Face or Arc'teryx, there are families for whom the woods of Grizedale exist not only for leisure but as an essential link to homes they had left behind and cultures they were trying to keep alive. My essay asked, 'What makes a forest?' and explored the natural history of Grizedale Forest itself, inhabited overwhelmingly by the Sitka spruce species of tree, originally from the west coast of North America. The tree was introduced to the woodland in 1831 and is one of Britain's most important tree species, accounting for nearly 50 per cent of British commercial plantations.

An immigrant, from the shores of the Pacific.
Hardy, useful, now prolific.
In strength and height,
in width and might,
its forest siblings pale.

*If two trees make a forest,
then Sitka makes Grizedale.*

Uprooted, displaced, rehomed: the story of the Sitka spruce. A long way from home, integrating into a new community, called upon for its resilience and industriousness: becoming the backbone of a nation. The young used for paper, the old for ships – a vessel for words and journeys alike.

*Uprooted, displaced, rehomed; the story of immigrant families. From Ghana, Jamaica, Syria and Spain, some came for prosperity, some came to escape. Doctors, nurses, lawyers, cleaners. A long way from home, integrating into a new community, called upon for their resilience and industriousness: becoming the backbone of a nation.
As we look upon Grizedale, we look in the mirror.*

The living world knows no boundaries and provides solace for us in unrecognisable times or places. Yet, an experience so universal, that of connecting to the living world, is often policed with expectation and perceived superiority. In her book *Wanderland*, the Indian nature writer Jini Reddy shared her own feelings of inadequacy within the 'nature space', reflecting that she felt 'too conventional for the pagans, too esoteric for the hardcore wildlife tribe, not deep enough for the deep ecologists, not logical enough for the scientists, and not enough of a green thumb for the gardeners'. Early in my own journey as an environmentalist, I experienced similar feelings of inadequacy. My inability to name tree species, to identify them from a certain point in their leaves or grooves in their bark, was a source of immense shame. I was embarrassed to know the names of only less than a dozen birds and to be even worse at identifying their calls. I felt blind to the world around me and fearful that this made me less of an environmentalist, less of a nature lover. There is a popular Instagram post that regularly resurfaces showing on one side many logos of well-known brands and on the other a collection of leaves. The post laments the fact that most people would be able to identify all the logos but none of the plants from which the leaves came. And it's true. I couldn't. Yet, when I'd walk into a woodland, my heart would leap into my throat. When a bird, known or unknown to me, would sing within a thicket of trees or deep within a hedge, goosebumps would rise and pattern my skin. When my toes touched the lip of the sea, taking me deeper and deeper until my shoulders were under water, I would feel like God Himself was communicating with me. Were these experiences not also ways of knowing? There is a deep sadness and confusion that comes with experiencing an innate love for and desire to connect with the natural world yet feeling that this love and yearning must be expressed within the bounds of the rules of the nature 'elite'. The self-imposed bouncers, making sure only the 'right' people get tickets to the show that the natural world intended to be seen and experienced by *all*.

>*All.*
>
>   *Us all.*
>
>   *All of us.*
>
>   *All for us.*
>
>   *Or.*
>
>   *Just some of us?*

For a long time I did not know how to resolve these feelings, until I read the words of the author, poet and wildlife biologist J. Drew Lanham, who wrote so eloquently in his book *Sparrow Envy: Field Guide to Birds and Lesser Beasts* that, 'it's not so much about identifying what birds are, as feeling who birds are'. It was not until I began my research in Ghana that I experienced first-hand what this really meant, watching in awe as community members identified, described and told stories about birds that had no specific name in the Akan Twi language. Their knowledge is deep, innate, tacit. There, the mark of deep connection lies not in how well you can name the world around you but how well you *know* it.

The behaviour of gatekeeping nature is not particularly rare; most people of colour will have a story or two to tell of being reprimanded by an infuriated white person in natural landscapes, flabbergasted at the gall of someone who looks different for entering land deemed not for them. In 2020, we saw an incidence of this type of interaction, which occurred on the same day George Floyd was murdered, go viral. The Central Park birdwatching incident, or so it is called on Wikipedia, saw Christian Cooper (a Black birder) accosted by Amy Cooper (an unrelated white woman). What started out as a ramble in the park and a chance to do a spot of birdwatching for Christian, turned into an uncomfortable experience with Amy calling the police to inform them that 'there [was] an African American man in Central Park'. The unbridled racism exhibited by Amy was clear, and knowing the history of racism in Central Park, from its very inception which displaced the majority Black population of Seneca Village, made it even more exhausting. Fortunately, this time, justice was served, and Amy was charged with filing a misleading police report, although the charge was subsequently dropped after Amy attended an educational therapy programme. In an interview with NPR, Christian put simply what we all know to be true, 'The birds belong to no one.'[18] The cliffs belong to no one. The sea belongs to no one. The forest belongs to no one. And though we might like to trick ourselves with borders and deeds and legal ownership frameworks that some of us own and can use to prevent others from connecting to that land, this simple truth still holds. In the words of Ellen Miles, guerrilla gardener, author of the similarly titled book and inspiration for the title of this chapter, 'nature is a human right', and we must *all* be able to enact that right wherever and whoever we are in the world.

Like our siblings across the Atlantic, many people of colour in the UK have felt unable to explore, despite their desires, the wonder and beauty of the British countryside, many scarred by terrible, heart-breaking experiences of racism. Although there have been well-meaning efforts from established British nature charities and organisations to 'increase diversity' and attract marginalised communities to the natural world, they often lack the humility and an eagerness to really know and understand these communities. My own experience, working within a nature charity, allowed me to see how internal prejudices, fear of losing already established members who are often racist, and an inability to move past seeing race through tokenistic lenses, hinder true progress and inclusivity. In recent years, this narrative has been rewritten. When I spoke with Mya-Rose Craig, commonly known as BirdGirl, we reflected on her and her family's collective process of healing through birdwatching and how those experiences led her to advocate for diversity in the outdoors. In her book also titled *Birdgirl*, she detailed her mum's struggles with bipolar disorder and how the adventures in forests around the world, monitoring and later helping to conserve endangered birds, held the family together. As we spoke, she recounted that it wasn't until she was writing that she realised just how fundamental those years of hiking, walking and birdwatching had been. Subconsciously or not, her experiences led her to become a staunch advocate and activist for diversity in the outdoors, founding the charity Black2Nature at just thirteen. The charity engages marginalised children and teenagers in environmental justice education through regular nature camps, walks, day trips and events. Like her, many others have come together to create their own safe spaces within which to explore the living world. In July 2023, I was lucky enough to go on my first Flock Together walk. I had followed the work of Nadeem Perera and Ollie Olanipekun for years, eagerly opening every newsletter detailing the arrangements for the next trip. I hadn't ever been able to match up the walk dates with my schedule; so, I was ecstatic to finally come together with them and a wonderful group of walkers for a birdwatching walk at the Royal Society for the Protection of Birds Rainham Marshes, just forty minutes east out of London. The joy of the day was captured in the heavy, happy exhaustion I felt on my way home. The feeling that your body and mind has done good work, has been refilled with the energy of the natural world and of others. The space that Flock Together has curated, where people of colour can feel free, excited, energised and educated within natural spaces they previously felt excluded from, is powerful and now exists all around the country in a growing collection of people of colour–led nature groups. From Black Girls Hike, an outdoor organisation, founded by Rhiane Fatinikun MBE, 'encouraging

Black women to reconnect with nature through nationwide group hikes, outdoor activity days, and training events', to Fruity Walks, founded by Divya Hariramani, a London National Park City Ranger, who connects Londoners to urban foraging by mapping fruit trees and exploring the stories behind how they arrived and the people that planted them, many groups are healing the relationship between people of colour and the natural world. These organisations embody what is essential to the process of healing our relationship with the natural world – that it is not only an individual pursuit, comprised of solitude and contemplation, but a collective practice.

In the next chapter, we will learn that healing does not come from beautiful, bucolic experiences in nature alone. Healing can be gnarly and fearsome. It is a task that requires us to collectively reckon with our pasts in order to build our future together. It requires us to be brave enough to see the cracks and move, hand in hand, towards them.

26 Healing Through the Cracks

Healing is not neutral. This is an uncomfortable but necessary truth that we must not shy away from addressing. We cannot talk about reconnecting with the living world, creating more access or conserving precious ecosystems without understanding the systemic forces that have disconnected us and continue to do so in varying ways. Our collective pasts hold horror and healing in equal measure, and in order to commit fully to reconnection to the living world, we must peer through and enter the cracks, the uncomfortable spaces, to emerge renewed. In her 2019 essay 'Touching the Earth', bell hooks shared her belief that when we love the Earth, we can love ourselves more fully. She tells us that she knows this is true because her ancestors taught her it was so. Yet, so many of us are disconnected not only from the ways our ancestors communed with the living world, but the living world itself. Some of our ancestors took away land forcibly and destroyed it as a show of dominance. And over the last hundred years or so, generations of us have experienced years of intense physical and philosophical disconnection from the natural world, fuelled by urbanisation, modernisation, colonialism and technological development. These systems, explicitly or not, teach us that the natural world is beneath us. That it exists to be exploited by us. That,

Nature is a resource.

Nature is capital.

Nature is the periphery.

For those of us brought up under the influence of Western philosophy or religion, the impacts of this disconnection on our bodies, our minds and our hearts are oftentimes imperceptible. There's a quiet nagging, an imperceptible whisper just out of earshot, calling us to witness our collective grief. But, for the land itself, and the people whose cultures and lineages are deeply tied to it, the trauma is visceral. In her article 'Climate Anxiety: An Illness of the System', climate campaigner Ayisha Siddiqa highlights the distinctly different experience of eco-anxiety between those living in the Global North, often experiencing pre-traumatic eco-anxiety, and those in the Global South, who experience post-traumatic eco-anxiety. This knowledge aligns with the observations of many researchers who find that 'climate anxiety concerns predominantly [the] white middle-class, who dread the dystopian future that many unprivileged communities have already been living through for decades or centuries'.[1] Indigenous, Black, Brown, Pacific and East Asian communities, amongst many, are suffering the ongoing repercussions of enslavement, colonisation and imperialism. Living in areas that are overwhelmingly devoid of nature and overwhelmingly plagued by pollution. Witnessing their ancestral lands acquired for oil production, property development, industrial farming, mining and waste incineration. Or desperately wanting to reconnect with their traditional lands that they are now barred from entering without paying a fee. For many of these communities, healing is not the process of learning to respect and reconnect with the natural world, but a reckoning with deep feelings of solastalgia. Theorised by Australian philosopher Glenn Albrecht, solastalgia is defined as 'the distress that is produced by environmental change impacting on people while they are directly connected to their home environment'.[2] From communities in Northern Ghana, watching as their land transforms into desert – a process ever accelerating due to the climate crisis – to the communities affected by hurricanes and oil spills off the Gulf of Mexico. Solastalgia acknowledges the detrimental psychological effects of 'chronic environmental destruction' on and in landscapes one has always called home.

When we trace the roots of this disconnection – past the petrochemical and mining companies, past the 'evil villains' who we easily, and rightly, point our fingers at – we begin to understand that there are deeper philosophical and religious schools of thought and acts that lie at the heart of our eco-spiritual detachment and the overwhelming disconnection of marginalised communities from the land.

Take up the White Man's burden –
Send forth the best ye breed –
Go bind your sons to exile
To serve your captives' need;
To wait in heavy harness
On fluttered folk and wild –
Your new-caught sullen peoples,
Half devil and half child.

> *Take up the White Man's burden –*
> *Ye dare not stoop to less –*
> *Nor call too loud on Freedom*
> *To cloak your weariness;*
> *By all ye cry or whisper,*
> *By all ye leave or do,*
> *The silent sullen peoples*
> *Shall weigh your Gods and you.*[3]

These grim and uncomfortable words are taken from a longer poem, 'The White Man's Burden' authored by Rudyard Kipling, famously known for writing the collection of stories *The Jungle Book*. In these words, we get a glimpse of the doctrine that permeated the minds of colonists and motivated their exploitation of lands across the Global South. This doctrine has been referred to as 'Three C's of Colonialism: Civilization, Christianity, and Commerce'. The first and last, civilisation and commerce, are well-known motivations for colonisation. Europeans saw Africans, Indians, Native Americans and Aboriginal communities as inferior, not even human. It was their job, their burden, to civilise them. And given those of the Global South were 'so incapable', it was seen as essential for 'the White Man' to make use of the natural resources that lay in abundance, unexploited by the 'sullen peoples', to reap the infinite profit that lay in the land – the theory that forms the foundation of our extractive capitalist society today. But, what of Christianity? How did the religion that gave us the commandment 'love thy neighbour' lie at the centre of a cruel regime? Speaking with Obatala Efunwale, a traditional Nigerian spiritual practitioner and teacher, one possible answer emerges. He tells me that 'Once religion becomes a means of justifying business, it becomes industrialised, moving away from what religion is supposed to represent.' Europeans, from the French to the British and the Dutch, saw Christianity as a hallmark of civilisation and their own superiority – 'Unfamiliar with the diverse cultures on the continent of Africa, European explorers viewed practices unfamiliar to them as lesser and savage.'[4] They were motivated by the call of divinity, of heeding the word of and becoming closer to God. Often, it seems, they simply wanted to play God. Consider the decimation of Indigenous populations in North America upon European arrival, the hundred-year period known as the 'Great Dying', where nearly 60 million Native Americans were killed, which was seen as 'a divine act, clearing the way for European colonisation of tropical lands'.[5, 6] Even those most pious in the colonial era, those who believed themselves to be conducting good work under the word of God as missionaries, perpetuated the harmful and oppressive behaviours that often ran 'ahead of empire'.[7] Missionaries, especially Jesuits, functioned as explorers and meteorologists as well as evangelists, and provided some of the earliest reports on distant environments and peoples.[8] Documenting the harshness or profitability of environments and lifting those of the tropics into the superiority of the Lord's word; severing ties forever to local traditions and spiritual practices. By degrading, demeaning and disappearing the spiritual, cultural and earthbound practices of the people they sought to conquer – their *new-caught sullen peoples, half devil and half child* – European colonists embarked on converting Indigenous communities the world over, supplanting their beliefs with the Christian faith. Despite the rise of secularism in the West, our dominant practices and values continue to be informed by teachings of Christianity as well as those of Western philosophers such as Descartes. In Genesis 1:26-28, the reader is told that God created mankind to be 'fruitful and increase in number; fill the earth and subdue it' and in his famous 1637 paper 'Discourse and Method', Descartes proclaimed that 'there is nothing which distances feeble minds from the right road of virtue more readily than to imagine that the soul of animals is the same nature as our own'; the overall sentiment being that man and nature are separate and that nature exists as an inferior subject to be exploited by man.[9] These philosophies and doctrines have had profound effects on our (dis)connection with the living world, with our continued acts trying to establish superiority over all other life and our collective perpetuation of land degradation resulting in our physical and mental suffering.

For many people within and outside of the mainstream environmental movement, the knowledge of Christianity's extractive and colonial role in planetary and social exploitation is good enough reason to discount the practices and teachings of faith from environmentalism. When it comes to religion and the climate crisis, for many, the contradictions seem too stark, the damage too severe, and so a deep and ardent atheism is often a badge worn proudly by the committed environmentalist. Yet so much of today's environmentalism, so divorced from spiritual practice that moments of introspection and reflection go overlooked, still keeps alive the doctrines of Christianity and Western philosophy in its approaches to healing the planet. Our approach remains rational and logical. We import our ideas and solutions from the Global North to the Global

South, deeming marginalised nations and peoples as less capable of regenerating the land. However, while the US and UK lag behind on renewables which account for only 13 per cent and 35 per cent of their respective energy mixes, countries like Kenya steam ahead with 80 per cent of their energy produced from renewable sources.[10] We work on cycles of burnout, fuelled by the capitalist obsession with productivity, at the expense of care, community and connection. We act out of a sense of superiority, proud to be doing more and better than others, disappointed at their perceived lack of industry. Between the shaming, the blaming, the pointing, the dictating, when do we leave time for the noticing, the revering, the worshipping? So engrained in the environmentalist psyche is the notion that religion puts individuals on the 'wrong side of history' that when long-serving, world-changing activists, like Vanessa Nakate, share their faith on social media, they are met with collective ire. Taken down and 'cancelled', because faith doesn't belong in our attempts to heal the world and ourselves.

On the eve of Earth Day in 2024, me and my dear friends and co-organisers Imogen Malpas and Marie-Therese Png along with four musicians and facilitators, Dan Gorelick, Alexandra Yellop, Mantawoman and Hannah Yu-Pearson, brought fifty people together in the first iteration of Earth Church, a non-denominationl, multi-religious space of collective healing. After years of speaking to both Imogen and Marie-Therese about the lack of spiritual practice within the environmental space, the dearth of spaces for collective healing, grieving and imagining, our dreams of creating a space that offers a sacred moment of reflection for ourselves and our community came to life. In a world of militarised capitalism and hyperconsumption, alienation from the self, community and nature has become the norm. With the event, we aimed to reclaim worship as resistance as well as to offer the experience of church as a 'third place', which has been lost to many – especially young people. We wanted to help people move past the trauma so commonly held in relation to religion and reconnect instead to the divinity inherent in the living world. We hoped to build capacity for the collective strength needed to continue the fight for liberation through re-nourishment, communal healing, joy and celebration. The day-long service, which included a tea ceremony, impromptu musical compositions and musical performances on the yangqin and cello, a deep listening session with sound recordings from Ghana, poetry and altar-building, was held at Abney Park Chapel in Hackney, London, the first non-denominational chapel in Europe and the resting place of hundreds of activists, radicals and dissenters. The day gave us all the chance to settle into worship and faith nurturing practices of awe and reflection, of honouring the world as sacred.

Resisting the common urge to engage in environmentalism as an act of avoidance, a way to pacify our fear and confusion, Earth Church was a space to honour and surrender to un-knowing, to the unknown. Is that not faith, basing our belief and actions on that which we cannot see but that we know touches our lives? As Daze Aghaji told me over dinner, 'We are too scared to stop, sit and observe, afraid to see the darkest parts of ourselves.' If we stop and introspect, our images of superiority and selflessness might become blurred and cracked. But it is in these cracks, Báyò Akómoláfé tells me, that we must enter. They are portals, to understand the work that must be done. He tells me that 'healing needs the crack, the wound, the opening, in order to do new things and climate chaos is the proliferation of openings, is a proliferation of cracks'. Instead of entering these spaces of possibility, we paint over them and police them. We demean spiritual practices that look deep inside them and instead, tell the world that, though the cracks exist, we can paint over them with large-scale technologies, carbon credits, greenwashing and oppressive conservation practices. 'There seems to be something conservative about popular accounts of healing', Báyò reflects. 'When we speak about restoration and repatriation, it often seeks to restore the image of the familiar', business as usual. In this sense, healing becomes problematic, it becomes political. Business as usual is serving neither the planet or its people and the unabated continuation of colonial and oppressive practices in our current approach to environmentalism 'start to become carceral, an imprisonment of some kind'.

Around me lay filth, excrement, pain and suffering.
In my ears rung the screams of the children,
the shrieks of the women and the drones of the dying.

The chains rattled, as they pinned us tightly to
the damp wood of the hold.
We were on our way to misery, but the misery
had already arrived.

Through the wood I spied a crack, and through it,
a crashing display of white foam and indigo wave.
The water beckoned me. It licked at the cracks.
It was breaking open a portal, to come and rescue us.

HEALING THROUGH THE CRACKS

I would have jumped over the side, but I could not.
The crew has erected nettings to keep us safe.
They watch us closely, lest we should leap into the water
And I have seen some of my fellow prisoners
Most severely cut for attempting to do so.[11]

I wrote this fictionalised story from the perspective of an enslaved man aboard a slave ship journeying to the Americas using accounts detailed in an article published by the Royal Museums Greenwich and archival material of the accounts of the Igbo writer Olaudah Equiano, first published in London in 1789. Báyò and I reflect on what imposed safety would mean in the context of a slave ship. How for the slave owner it secured profit and capital, whilst condemning the enslaved to pain and suffering. When healing becomes containment, the infrastructure of the status quo, it becomes problematic, it becomes paradoxical. The antidote then, as Báyò told me, is fugitivity, is maroonage, departure. To run from the forces of oppression, and the false sense of security they advertise, towards the cracks.

Healing is a process not without grief. We will lose things and become lost ourselves. But we cannot transform our world without accepting and engaging with this grief. Honouring what will or can no longer be and fighting to protect what remains and is coming. In the next chapter, we will draw near to grief. We will learn from those experienced with its darkness and magic, and its role in ushering new worlds.

27 Grief Is the Way to Transformation

In early April 2023, I returned to Ghana to continue my research. I was excited and full of anticipation. Memories of the sheer beauty of the landscape I would return to and the people I would reconnect with flooded my mind. Turning from the main road onto the red clay track that led deeper into the forest, my heart began to soar, the excitement building. I looked out of the backseat window, eager to soak up the various shades of green that painted the landscape. As we drove – my trusty field team member Ben, a wildlife expert, Enock, my research assistant and our wonderful driver, Mr Nkrumah – we waved and greeted passing community members carrying large bowls made of hammered tin on their heads and gleaming cutlasses in hand, off for a day's work in the surrounding cocoa farms. As the car approached the village, less than 100 metres away from its entrance, my heart sank and the mood in the car soured. To our right, where once stood a variety of native trees and locally owned cocoa farms, lay a cavernous hole. A hole so large and striking that only a fleet of industrial quarrying diggers could be responsible for it. The hole was filled with murky brown water, rerouted from the local stream, sludge running down the sides of the banks of what now resembled an emptying lake. It was shocking to revisit a place I had so deeply connected with only a year prior, now changed forever. The mines – plural, for a second went under construction as my work continued – lay frighteningly close to the village school where community members' children spent most of their days. My heart was in my throat, and I felt a stinging in my eyes. What we witnessed was an illegal gold mine, locally referred to as *galamsey*. Something that almost every Ghanaian is familiar with and that burdens the country. But this wasn't a time to linger. Even though I wanted to run towards the hole, to document the damage and grieve the destruction, loitering any longer than we already had would be foolish; mines are well-defended sites of hostility, both ecologically and sociologically. That first day I was numb, but the following morning I was overwhelmed with an intense sadness. I spent the first three hours after waking sobbing deeply – I couldn't even hold it together for the meetings with my supervisors I had scheduled. The sadness quickly grew into anger. My despair became rage – hot, angry tears scorched my skin in frustration. Worse still, there was no enemy I could direct this anger at. Illegal gold mining in Ghana is not directly conducted by big corporations on the ground. There is a chain that runs from large corporations, to politicians, to local communities that keeps the whole system anonymous and elusive. Those carrying out the gruelling work within the mines were motivated by the prospect of building a better life for themselves and their families. My heart ached. 'Why is the world like this?' I screamed in the void of my mind, 'Why are we like this?' For a day or so, I felt helpless. People often ask me how I stay hopeful about solving the environmental crisis, how I am not frozen by the catastrophe we are witnessing. In that moment, I was void of hope. I was exhausted, tired and exasperated. I was grieving. Yet, days later, I was back in the forest, laughing and connecting with community members, pressing on with the work I had set out to do. Not because I had forgotten about the mine or was ignoring it, but for the simple fact that being derailed was not an option. It was not an option for me and my research, but more importantly, it was also not an option for the community members, who worked every day to sustain themselves and their families, who would continue to live on and love the land for generations to come.

My instinct was to get out and organise, to occupy the mine, to do something. But facing violence and even death would be the inevitable fate of those who might be brave enough to confront the miners. I decided to decentre myself and to listen to what the community members had to say. In the small wooden church that was the venue for all our meetings, we spoke for two hours straight. They spoke, I listened. They told me about the structures that make these mines possible. They shared their heartache and mourned what was lost, but they also shared words of hope, of wisdom. They shared proverbs and teachings that rooted them in place and in mind.

An event which seemed to trigger intense reactions within me was approached by the community through dialogue, reflection and deeper connection with each other and the land they called home. I was learning first-hand how reductionist Western environmental action can sometimes be. Searching for a single enemy, a single solution to the destruction that frontline communities, like those I was working with, face every day. The community members were not without anger and frustration, but neither did they surrender to the grip of despair. What other option did they have? There was water to fetch, children to teach, land to tend, lives to live. There was work to be done, and it had to be done because that work was good, it was necessary, it was essential.

Earlier in this book, we explored the role of rage in

rooting our environmentalism, in preparing a way to find liberation in times of destruction. Yet, we often centre rage, in all its diversity, as a core route to connection. We want to witness rage in others, and feel it within ourselves, to measure our capacity to care. We wait, poised, to assess how quickly we and others react to distressing news and use the intensity of that reaction as a proxy for morality. But what I learned in the small wooden church that day, I felt deeply in my bones. That it is not always rage but often connection and healing that we need. Our Western culture is practically void of spaces of communal healing or practices of communal grieving. Further, many of us are untethered from our own individual practices of contemplation, religion or spirituality that we can draw on in times of despair.

Upon returning home, I continued to reflect on this experience. On how I was enabled by the practices of the people I was collaborating with to connect deeper with the natural world around me. To drink from crystal clear waterfalls, scramble up vertical inclines of the mountains, lizard like, to pause, stop and listen to every chirp, tweet and song resounding from the treetops. We never stopped talking about the mine, yet we never lost our awe for the world around us and the relationships we were building. We funnelled our grief into conversations, brainstorming, planning, drawing, hiking, recording and collecting. It wasn't until later that year, as I listened to an episode of the *For the Wild* podcast with Yuria Celidwen, Báyò Akómoláfé and Naomi Klein on climate grief and hope, that I was able to better place my experience. During the conversation, it was the following words from Yuria Celidwen – an Indigenous researcher born into a family of mystics, healers, poets and explorers from the highlands of Chiapas, Mexico – that found me.¹

'Grief is the way to transformation.'

These few words hold the simple yet powerful truth that in endings lie beginnings. That death is birth. That through grief lies the way to hope.

As my husband and I walked along the River Menalhyl in north Cornwall with the forager and film-maker Seth Hughes, he shared, generously, his own story of grief and how through the living world his pain was transformed into healing. Seth, like most young white men in the UK, had no connection to a spiritual practice; he admitted that so much of the way spirituality had been advertised or practised, and even the language used, was a turn off – an experience and feeling common to many. But when his father suddenly passed away, he looked inward and after ten months of reflection, of being held by the grooves of the rugged Cornish coast and the canopies of its gleaming green woodlands, he felt that an organic spirituality was let out of him. As someone who works to reconnect the British public with the spiritual teachings of the land, to reconnect with traditional foodways and land practices, he made grief the catalyst to prompt his curiosity about his lineage and how it was tied to the living world; and grief continues to inform his environmental practice today. 'You have to work out what you're grieving', he tells me as we skip over large muddy puddles and rivulets. 'I felt this grief in me my whole life, but I wasn't aware of it until my dad died; it made me look back and connect the dots and to unlock other grief that I had to get over.' Grief about his lack of connection to his ancestors and grief about the loss of life on Earth. In the Western world, loss as drastic as Seth experienced, let alone such life-altering grief, is something deeply uncomfortable, cast to the shadows for the individual to either mask or hold alone. Max Porter writes wonderfully and honestly in his book *Grief Is a Thing with Feathers* of the oft felt loneliness experienced by those grieving and how they are left mystified by 'the disjunction between the vastness of their catastrophe and the world's muted response'. Yet, in Ghana, grief is something plastered on billboards – literally. Saturday is funeral day, and funerals are attended not only by close family, but communities, villages, towns and even strangers. Driving down the motorway, every 20 metres or so, you will see a large billboard, with affectionate words about the deceased 'wonder mum', 'loving father', 'beloved aunt', accompanied by a large-scale image of them and logistical details about the dates and times of the various ceremonies and upcoming events to mark their death. For those within the Ga tribe, death is met with generations-old creative traditions. Steeped in cultural and religious practices and the belief that life transcends death and that the deceased will continue with his or her profession in the afterlife, they are often buried in a fantasy coffin, a representation of their career and livelihood, allowing them to remember where they came from when they pass into the next life.[2] Many of the carpenters who craft these coffins start their journeys when they are as young as fifteen and there are only around eight or ten practising carpenters in Ghana today. Their art conforms not to a preconceived standard of creativity, but rather grows as a manifestation of the culture of their people, providing secure livelihoods for generations of practitioners. There are plenty of other cultures and traditions that approach death in a myriad of ways that are connected to collective practices of grieving. Yet, there is so much about grief and the path to transformation it reveals, within the natural world alone.

In her book *The Bleeding Tree*, writer and folklorist Hollie Starling moves through the pagan 'Wheel of the Year', recounting her experience of suddenly losing her father. She reminds us how 'ancient communities understood the flip side of the symbol of death', that it holds the potential for new life. This is a lesson embedded in the threads of the natural world, famously in the life cycle of the magnificent yew tree. In the Celtic tradition, the yew represents death and resurrection. Not only do yew branches root themselves in the soil below, planting themselves to create new trees, but at the end of a yew tree's life, a show of powerful surrender occurs.[3] As a yew tree dies, it gradually hollows out from the heartwood, becoming a shell of its former self. During this process, it produces new roots from its cavernous centre, stabilising itself as it transforms into a home for new life, allowing its young to grow from within.[4] What if we approached environmental healing and action as a yew approaches death? Slowly but surely removing the systems that no longer serve us, hollowing out to make room for new worlds. What if, by rooting our environmentalism in healing, following nature's lead, we see that our work is about letting go of the dying parts of the present to provide the scaffolding and space for the future?

On a sweltering September afternoon, I had the pleasure of speaking with Willow Defebaugh, who is the co-founder and editor-in-chief of *Atmos*, a beautiful biannual magazine exploring climate and culture. Speaking candidly about their transition and what that experience had taught them about grief and transformation, they shared an essential process that our cultural stories often neglect when documenting transformation in nature and within our own lives – decomposition. We settled into a conversation on the age-old story of the caterpillar and the 'heroic ideas of what happens to it within the chrysalis'. Yet, within those translucent walls, the caterpillar ceases to exist, its whole body decomposes, becoming liquid with only the imaginal discs, cells that have existed since the larva stage and will create the adult body, remaining. The transformation of a caterpillar into the beautiful butterfly we admire is one of duality, 'where there's nothing but decay and loss within that little space, and at the same time, you have the blueprints, that were always there, for what it was going to become'. Willow has gone on to create the Chrysalis Youth Fund, an initiative to transform and support youth climate activists around the world.

As we watch the Earth burn, flood and break, grieving all that has been or will be lost, how do we cultivate an understanding and acceptance of this transformation? That held within the multitude of disasters, the inevitable breakdown of extractive and oppressive systems, and the myriad personal shifts we will need to make, lie the blueprints for a renewed world. What tools, systems or philosophies can we hold on to and be led by to process our eco-grief and begin the process of creation, of rebirth?

In the March of 2018, a group of African Americans and Black diaspora gathered in a forest near Roswell, Georgia, to mourn the lynching of Mack Brown, which occurred in that same place in 1936. They gave offerings to the river, the place where Brown was 'grossly interred by the men who murdered him', venerating his life and the life of the water.[5] In the embrace of a nearby grove, under the witness of a magnolia tree, they settled into a circle and wept. They cried and prayed and sang and read. They 'meditated on the ways white supremacy used the Earth as a surrogate to do harm against Black folks. [How] the trees, soils, rivers, mountains and minerals wanted nothing to do with their violence'.[6] They were guided to this place by the folks from the Equal Justice Initiative – the organisation behind the Community Remembrance project that provides commemorative and archival justice for the descendants of racial terror lynching victims to honour the unjustly departed members of the African American community. Following in the tradition of 'swords to ploughshares' – the act, of Biblical origin, that sees weapons turned into objects of civility – and by the Mexican project Palas por Pistolas, which enabled citizens to hand in weapons in exchange for domestic appliances, they used recycled guns, transformed into trowels, to dig and preserve the soil of the land.[7] The guns were melted and transformed into trowels by RawTools, an organisation that, as well as turning weapons into tools, offers non-violence training and community peace-building resources. Some of the soil would return, in a labelled jar, to the Equal Justice Initiative to be preserved in the archive. The rest was wet with the water from the river and would nourish the fifty trees that they would go on to plant across Atlanta on 4 March 2018, the anniversary of Dr Martin Luther King's assassination. After the soil ceremony, they were led into a deep nap with the author and founder of the Nap Ministry, Tricia Hersey, and gave their 'rest as an offering for the safe passage of [their] unjustly departed ancestors, rest taken from our enslaved ancestors . . . the rest taken from the earth'.[8]

This ceremony was led by Lead to Life, a collective led by Black-diasporic and queer artists, healers and ecologists. Lead to Life bridges racial and environmental justice through ceremony and art practice to cultivate a time where Black diasporic communities and environments are free from the violence of white supremacy and environmental desecration. I had the honour of sitting with brontë velez, Lead to Life's founder and creative director, and hear them reflecting on the need for spaces of communal mourning. For them, the act of planting trees, of nurturing the land while grieving with it was a powerful act of witnessing. They ask me to 'imagine the gun as a descendant of rock and soil, mineral and wood, and think about the ways that trees have become viable means to harm Black communities'; the thought is gut-wrenching – there is so much to grieve in the transformation of sacred earth into violent weapons. Without spaces where we might join our palms and tears to express this grief, it stays trapped and turns into stored pain. As brontë shares, accepting what is lost is radical, and we must be 'in community with the pain, letting the pain impact us is important'. For brontë, 'grief allows us to be in contact with what is happening with the Earth, what has happened to our ancestors without needing reconciliation'. Just as the Black community cannot recover our lost ancestors, the lost Black children, the lost Black bodies, there is much on this planet that we will never get back. As we fight, as much as possible, the myriad environmental crises facing our planet, we must also carve out space to be transformed by our grief.

28 From Grief to Care

Regardless of whether we have accepted it or not, each one of us, in our own ways, is grieving the ongoing destruction of the living world. We send our emails, we grieve. We walk to work, we grieve. We bask in the sun, we grieve. Most of our grieving is imperceptible, muffled under our conviction that either all is lost or that someone is coming to save us or that if worst comes to worst, our privileges can save us. We neglect our grief, catching glimpses of it through our day, but never looking it straight in the eye. So much of what makes facing and embracing the grief we hold – as we witness the suffering of the Earth and communities across it – is our lack of care infrastructure. We can never be sure whether our grief will be held or rejected.

As I edit this section in the July of 2024, during a week of fear and anxiety, witnessing vile racial violence take hold of the UK, it has been the people, not those in power or the police, that have brought me hope and feelings of safety. Witnessing thousands of allies of all ages and backgrounds come out to resist hatred and keep their communities safe, it is clearer than ever that our radical practices of care, if anything, will be what saves us. In the last and final section, CARE, we will come to see that *we* are all *we* have, and that our survival will be rooted in the ways we show up for each other and the living world.

Care

29 We Are All We Have

A week before my eighteenth birthday, I left my family home. When I talk about this tumultuous time in my life, people often tell me it was a brave thing to do. At the time, it didn't feel brave. It felt terrifying, world-ending and urgent. It was an act that, literally, saved my life. I left my family home and was taken in by another. They weren't relatives or close friends – taking all things into account, they were strangers. A couple of months earlier, whilst stacking the shelves of my local Waitrose, seeds of a friendship that would change the course of my life were planted. We bonded over our love of Cat Stevens, environmentalism and politics, excitedly chatting as we straightened jars of tinned tomatoes. We met up one or two more times that summer, and for some reason my darkest troubles emancipated themselves from the chains of my mind and fell on her compassionate ears. On her eighteenth birthday, my new friend and her family welcomed me into their home. I remember her mum, my other-mother, describing me at the time as a 'petrified young woman'. I was gaunt and fearful and shy and withdrawn. But in the warm embrace of that home and in the swelling hearts of the teachers, colleagues, friends and strangers that surrounded me like a protective covering, all the broken pieces of me were collectively gathered and put back together again. This was my first experience of collective care, of what the disability justice activist Leah Lakshmi Piepzna-Samarasinha calls a 'care-web'.[1] Of being nurtured emotionally, spiritually and economically by a wide network of generous beings who cared not out of obligation but out of love. It was a time where my understanding of family and community was flipped upside down. I was taught unconditional love, persistence and repair by my mother and sisters and the ways in which we were split apart and brought firmly back together. I was taught vulnerability, interdependence and acceptance by my other family. And I was taught trust and openness by all the strangers who cared, without knowing me very well, because they were simply compelled to.

The care, nurturing and nourishment I received in that frightening and transformative time were experiences I grew up being unable to imagine. Not as a result of my personal upbringing – my mother remains one of the most caring, selfless people I know – but rather as a result of the Western obsession with individualism, resilience, wellness and competitive self-enhancement. The obsession of *making it on your own* and *by any means possible*. I had deeply engrained notions of self-sufficiency and saw the need for help as a sign of personal failure. My experiences in later adolescence quickly shattered these fickle and flawed ideals. My survival became dependent on a wide network of care and mutual aid that filled in the gaps that our supposed social care system could not. With each layer of individualism I shed, emerged a clearer image of the invisible infrastructures of care that hold together life as we know it. Infrastructures that care scholars such as Carol Gilligan, Virginia Held and Nel Noddings have been trying to illuminate since the early 1980s.[2] Infrastructures that are aggressively devalued in Western society. Much of this devaluation arises from the perception of care as being gendered, weak and unproductive within our dominant system. A system where capital is king, profit is power, and the protection of and unending servitude to the economy is key. Yet, we forget that the economy is an iceberg, with the monetised economy propped up by ecological processes and care activities that lie steadfast beneath the surface. Modern economics treats negative effects on nature as externalities, failing to acknowledge the fundamental role nature plays for all economic production processes. Additionally, caring activities, commonly conducted by women and usually unpaid, create the social foundations that enable industrial economic activity but are diminished, not seen as real work. Our social and ecological interdependencies are devalued, and autonomy and independence venerated. Care work – interpersonal, intergenerational, interspecies, paid and unpaid – dictates the quality and longevity of all life on this planet. Yet, to the detriment of our communities and environments, it remains ghost work. Currently, 22 per cent of the UK population lives in poverty and out of that population, nearly 30 per cent are children.[3] In their 2023 end-of-year report, the Trussell Trust revealed that three million people had relied on food banks that year, a 37 per cent increase from the previous period.[4] Maternal care and support have significantly deteriorated and the growing elderly population are receiving less and less care.[5] Over 400 NHS staff in England are leaving their jobs every week due to burnout and poor working conditions, and hundreds of thousands of jobs remain unfilled due to undesirability and discriminatory immigration policies that stop overseas workers taking up vacancies.[6] As I write this, the Conservative government has brought in new measures, including removing access to free medical prescriptions or legal aid for those on Universal Credit, that undermine and threaten the quality of life for already vulnerable communities.[7] These measures – read punishments – are justified as a response to 'indiscretions' by those receiving social support such as missing job interviews or losing pay, ignoring that these actions often stem from limitations imposed by an individual's lack of resources, such as relying on inconsistent public transport, debilitating mental or physical illnesses that flare-up in unpredictable ways or lack of accessible or affordable childcare to facilitate looking for or taking up new work.[8] Whilst the important, rational and masculine work of decision- and money-making unfolds in the public sphere, caring activities are pushed behind closed doors, relegated to the private and individual domain.

As our circles of care shrink, we become more fearful or suspicious of others, and more interested in 'looking after our own', rendering our entanglements with our wider community and the rest of nature non-existent at worst and tolerable chores at best, or proof of a life protected by the shield of privilege.[9] Collective care has been branded a burden, replaced with the encouragement of and growing obsession with *self-care*. Not the radical self-care introduced to us by feminist scholars as an organising strategy to build community resilience or, as Audre Lorde presented, a political 'act of self-preservation'.[10, 11] But a *selfish*, self-*centred* type of care. One which has been co-opted by capitalism, perpetuating the idea that to be worthy of care you need to be able to pay for it. Capitalism-driven individualism sees care, much like natural resources, as a commodity. And it is no wonder that the main beneficiaries of this commodified care are those whose resources allow them to pay for myriad forms of support.[12] Experiencing an organic, mutualistic care infrastructure in my late teens was an embodied lesson of the politics of care. The memories of that time continue to illuminate the ways in which my improved living circumstances – which wouldn't have been possible without that support system – provide me with more access to safety, security and well-being. For me, it is impossible to ignore the fact that my individual well-being sits within a wider web of collective safety and prosperity. My experiences compel me to see self-care as an opening for joy, reflection and relaxation to re-energise our souls for collective action. This isn't to say that we don't deserve rest unless it is countered by an equal action: This is not a zero-sum game or tit-for-tat; we all deserve rest whether we have been individually or collectively 'productive'. But when our rest and care practices incentivise disengagement, sever the links between ourselves and the world around us and reinforce disconnection, it is transformed into neglect. A neglect that when compounded results in individual and collective suffering. We forget that, with each strand of interdependence severed, our collective existence and the health of the planet become more deeply threatened. We see this suffering in the depleted soil cover and rapidly decreasing yields caused by our disconnection from our food. We see this suffering in the dunes of discarded fast fashion items that litter cities and deserts, and lead to the exploitation of largely Black and Brown women in the Global South. As I write, the world is waking up to the plight of over six million people who have been killed and the further six million who have been displaced in the Democratic Republic of Congo.[13] They are the victims of violence and war fuelled by the greed for resources, the country being a leading supplier of the coltan and copper that we use in our mobile phones and renewable technologies. Mothers losing their children as young as five to the mines.[14] Desperate to nurture their families with limited options, made ever narrower by the West's rampant consumerism. We are in a crisis of emotional and physical disconnection. Our isolation becomes ignorance, which in turn enables exploitation. And this is not an accident. We have been conditioned to see Black, Brown and Indigenous bodies as foreign, faraway and helpless. To see them as less than. To see them as the unfortunate but necessary labourers tasked with the upkeep of our increasingly 'developed' and disconnected lives. Capitalism relies on concealed connections. It compels us to become so obsessed with our own lives that we have little time, capacity or capability to *truly* care – beyond experiencing a detached horror and overwhelm – about our entanglements with all other beings on this Earth. Capitalism relies on the perpetuation of our *mass relational drift*. It rejoices at a Pangaea of connection being transformed into a scattering of islands where, isolated, we forget that our small world is but a node in an infinitely beautiful and complex network.

These islands of isolation, which separate us from each other and the natural world, lie at the root of the Anthropocene or rather act as the distributed network that powers it. The Anthropocene is the unofficial geological epoch that describes the last sixty years of accelerated climate and ecological breakdown caused by humans.[15] The term was coined by ecologist Eugene Stoermer in 2000 but was hugely popularised, becoming a buzzword in the environmental space, by the Nobel Prize–winning atmospheric chemist Paul Crutzen. Although it hasn't been confirmed as an official epoch, most scientists are of the opinion that the term Anthropocene is a strong reminder to (and a mobilising phrase for) the public to highlight that, as a species, we are having undeniable negative impacts on the environment at planetary scale, to the extent that a new geological epoch has begun. For many, the Anthropocene is evidence that we are a careless species. But who is 'we'? It is not the universal, seven billion 'we' that has caused this crisis. Nor is it that universal 'we' who are beholden to the individualism and disconnection that have caused it. It is all too easy to render the world evil, to see *all* humanity as inherently uncaring, and become further entrenched in the fallacy of the self. But if we were being more accurate, we would have to say that the Anthropocene is really the age of Western-influenced domination and industrialisation. An age built on and that unfolds through our Western values, ideals and worldviews. What is made clear by climate and environmental breakdown is not that those in the West are void of caring but that we care *differently*. In fact, the West, as a society, cares so deeply that we threaten our own lives. People must go cold or hungry to care for the economy. Extraction must accelerate to care for our culture and lifestyle. Nature must be subservient to care for our progress. There is so much that we – those of us living in the Global North – fiercely care for; it is only that our priorities lie somewhat (read majorly!) misplaced. Our care and collective nurturing, intentional or not, of systems of oppression is a stark reminder that a large capacity for care is built into our society, systems and government; our challenge is transforming *who* and *what* we care about and *how* we care for them. To allow ourselves to be depended on, as much as we depended on. To become intimate with all beings on this planet in a reciprocal dance and, as the writer Katherine May described to me, work towards blurring the lines and harsh borders between ourselves and the living world around us. To radically reimagine ourselves back into 'this enormous, complex, planetary system' that we are in no way separate from. To allow our bodies to become receptive to the whispers of the trees, the hum of the soil and the thumping hearts of the people around us.

In this section, we explore the work and teachings of

life-givers, caretakers, nurturers, nourishers and those who have an intimate understanding of how the politics of care, in all its stages, manifests in our lives and environmental movements. We will bring to the fore the synergies between mothering, growing, kinship and solidarity, foregrounding care work as a key, yet oft side-lined, contribution to the environmental movement. This section asks, as queried in *The Care Manifesto*, 'What would happen if we put care at the centre of life?'[16]

30 Beyond the Burden of Climate Care

Within the mainstream environmental movement there exists not only a narrow conception of what climate work looks like, but also worryingly, conceptions of the act of care itself that remain largely one-dimensional. Instead of seeing ourselves as we truly are, beings intertwined with the rest of the living world, we position ourselves as stewards. As the benevolent folk descending from our lofty positions to protect a weak and vulnerable nature. A story little different than the damsel-in-distress fairy tales many of us grew up reading and watching. A story that centres othering, ego, heroism and distance. A distanced care is one that imposes instead of listens. One that rushes apologies and demands quick forgiveness. One that centres the comfort and stability of the self. This distanced care is shaping the way we approach environmental action and the solutions we revere and invest in. We look at environmental action through the lens of hindsight, trade-offs and scalability. We create carbon offsets so we can retroactively absolve ourselves of polluting the planet. We weigh the trade-offs between our 'eco-friendly' practices and years of extraction, obsessed with the pursuit of the *fastest* most *scalable* solutions, which won't fundamentally change the way we interact with the living world. There is a resistance, even repulsion, to caring and nurturing for the 'rational' environmentalist, often the older white man who insists the climate and environmental crises can and must simply be solved through technological or logical means.

> *Caring is a woman's practice,*
> *after all.*
> *And that's all good and well,*
> *when you're rearing children.*
> *Or looking after the elderly.*
> *Or helping the sick and the lowly.*
> *But it has no place meddling with*
> *serious matters of scale like*
> *saving the world.*
> *This is a technical job.*
> *A man's job.*
> *Care has no place here.*
> *Take it back to your quiet corners.*
> *Where you knit and chatter,*
> *And cry.*
> *Take it away from this*
> *solemn place of work,*
> *into the forest,*
> *into the woodland,*
> *into the garden,*
> *Take it back into your bodies,*
> *into your heart.*
> *And leave it there.*

So convinced are they that we are not nature itself, these people forget that caring for other people is caring for nature and, of course, vice versa. No, instead, to care for the environment, as with caring for others, is to take on a sombre obligation. It is a burden. One that is either solved by patching up the problem with large-scale technologies or placing guilt on the shoulders of the individual, neglecting to interrogate the systemic issues of power, resource and relative responsibility.[1] This paternalistic judgement has most famously been made through the creation of the concept of 'carbon footprint', a phrase that has imperceptibly entered our cultural lexicon. Now rightly understood as a call for individuals to take necessary personal action to reduce their impact on the planet, many are still unaware that the phrase sat at the centre of a successful 2004 campaign led by the marketing masterminds at the ad firm Ogilvy and commissioned by the fossil fuel execs at British Petroleum.[2] The campaign called for the everyday citizen to head over to BP's website and calculate their carbon footprint, with the pledge from the fossil fuel giant to reduce their emissions – which, in 2004, sat at 606 million tonnes of CO_2 – by 4 million tonnes: the equivalent of two and a half days of their yearly emissions in 2004.[3] Not much has changed, in the last decade, apart from the weight of the fossil fuel industry's wallets. We continue to hear tall tales of the industry's commitment to investing in large-scale technological infrastructure – curiously not often wind or solar – that will drive down their emissions. All the while, the grief-stricken cries of those harmed by their existence – the communities, the wildlife, the land – go ignored. The fossil fuel industry has no interest in caring for people or planet, of course we all knew that, but it is not only in these spaces where nurturing is neglected. Even in spaces that enthusiastically and proudly wear the badge of fostering deep environmental care, the standard practices often enable the systems of oppression so fastidiously campaigned against. In Western environmental spaces, care is patriarchal, paternal and colonial, supplanted with saviourism. Humans are seen as nature protectors, with a duty to 'save' the world and the vulnerable communities suffering the effects of climate and environmental breakdown.[4] We rely on moral politics and messaging, convinced we are on the superior side once enough environmental boxes have been ticked. We hoard examples of our care, squeezing them up our sleeves and in our back pockets, where they remain, ever ready to be strung out like the handkerchiefs of a magician, exhibited to the world. This thinking reinforces the distancing from the rest of the living world we are trying to resist. It entrenches hierarchies, harmful power dynamics and takes us further from each other and the natural world.

Nowhere can we see the cascading impacts of colonial care more clearly than in the conservation movement, especially within forest landscapes which serve as a habitat for millions of distinct species, store 34 per cent of the planet's terrestrial carbon and are home to numerous Indigenous and local communities. As these ecosystems suffer the dual consequences of the climate and biodiversity crises, there has been a proliferation of conservation projects monitoring important species, informing policy decisions and implementing change on the ground. But behind the wildlife mascots, wild expeditions and binge-worthy documentaries with high production values, biodiversity conservation has a dark colonial history that saw Indigenous communities across the tropics treated not as humans but as flora and fauna. These communities were severed from their lands in the name of planetary protection, in the name of planetary care. It is estimated that up to ten million people have been displaced from half of the world's protected areas – all in the name of biodiversity conservation.[5] Even though we like to believe that this era lies firmly in the past, forest communities continue to be excluded from conservation research and action. The time of oppression through conservation is not over, and many modern large-scale conservation projects have marginalised or inflicted violence against Indigenous Forest peoples around the world. One example is that of the Embobut Forest peoples of Kenya: Since 2009, 2,000 homes of this community have been burnt down in an act of forced eviction, dispossessing members of their ancestral lands, in order to create forest plantations for European carbon offsets. And although it might be tempting to think of the harm caused by conservation projects as being 'isolated' incidents that happen on a small scale, there is a pervasive culture of racism and ecofascism within conservation spaces. Take Bernhard Grzimek, a German wildlife campaigner in East Africa, who once said, 'A national park must remain a primordial wilderness to be effective. No men, not even native ones, should live inside its borders.' Dated and violent perspectives such as these fuelled the destructive actions of the past and persist even today. Yet, it is the particularly Western binary worldview that separates humans from the living world. That creates an 'us' and a 'them'. The 'savage' and the 'saviour'. A worldview that cannot conceive of codependence or collaboration and that crowns itself, and its people, as the superior stewards of this Earth. The majority of ecological care on this planet is undertaken by Indigenous and local communities who care for and defend 75 per cent of land ecosystems and over 60 per cent of marine ecosystems.[6] Those at the frontlines of forest degradation and destruction, the Indigenous and local communities – who live with, are dependent on and care for 80 per cent of the world's biodiversity – are not only victims of the climate crisis and colonial conservation practices but are also essential leaders in forest protection.[7]

It was in 2021 that a major report by a UN agency recognised and documented in detail the efficacy of Indigenous conservation for the first time.[8] The study found that 59.7

million metric tons of carbon dioxide (CO_2) emissions have been avoided thanks to Indigenous forest conservation in Bolivia, Brazil and Colombia alone. That's the equivalent of taking roughly 12.6 million vehicles out of circulation for one year. Reports like these tell us what Indigenous and local communities have always known and have been trying to communicate for decades, that their existence is not adversarial to, and often essential for, ecological flourishing. What we see, through the case study of conservation, is that colonial care is really about control, control of people and the land. If we fail to dismantle this type of care, we will continue to live in a care*less* society that functions on the same oppressive systems that have led to the current destruction we are witnessing.

As Katherine May says, we can no longer 'de-tooth' the Earth by calling her Mother and treating her as such. Applauding her as the model of nurture and sacrifice. Of using, exhausting and underappreciating her only to feign awe and care, of the selfish kind. The kind that's scared, repulsed even, by the rage, the jagged edges, the power, the magic, the gruesome, grisly bits. The Earth, its mothers and its vulnerable folk are not passive, silent beings for care to be 'weaponised against... bestowed [upon them] by a superior', says Katherine. Fighting for ecological care whilst neglecting to deeply connect to those most keenly aware of the complexities of care – the frustration, the ambivalence, the solidarity, the joy – results not in radical change but a world where we 'still continue to exclude neurodivergent people [and] disabled people ... it will still be individualistic, it will still see mothers of young children and elderly people being isolated'. The only difference will be that it is a little bit greener. As Katherine told me, there is a surprising and disarming way in which 'care suddenly becomes the exercise of power'. We were reflecting on the period of her adult life when she was diagnosed with autism, and the patronising and insulting experiences she had within the environmental movement, a movement she had, up until that point, felt firmly a part of. She described the way in which activism, protest and resistance have become acts inextricable from an exclusionary type of sacrifice. Not the sacrifice of inconsequential privilege or the pain of deeply caring for the planet, but the sacrifice that competes, elbows, demeans and badgers. The sacrifice that says, 'Look, I'm better than you', 'Look, I've won'. The sacrifice that insists we must all become shells of ourselves, frazzled and shrimp-like, a pile of ash with the last ember burned out. As Katherine makes abundantly clear, this sacrificial competition is one that 'only the able-bodied and the neurotypical can win'. This is something that the disability justice activist Leah Lakshmi Piepzna-Samarasinha speaks about so candidly in her book *Care Work*, calling out the burnout culture within the environmental movement that disproportionately harms and excludes 'broke folks, parents and disabled folks'. She comments on the inaccessibility of mainstream activism that still walks 'ten-mile-long marches, run[s] workshops that urge people to "get out of their seats and move" and fails to include disabled organising strategies'.[9] If winning means destroying all the joyful, restful, contented parts of myself, the glint in my eye, my uncontrollable giggles, the rooting of my body to the earth beneath my feet, then I think I'd much prefer to lose. The competition is as much a distraction as it is a fallacy. A distraction from the fact that the most vulnerable and dependent on various forms of care, especially disabled folk, hold a myriad of crucial survival strategies, essential for crafting resilience amidst crises. Those with chemical injuries, severe asthma and autoimmune conditions suffer disproportionately the impacts of air pollution and know what it is like for breathing to be unsafe. As Lakshmi Piepzna-Samarasinha reminds us, they know about 'masks, detox herbs, air purifiers and somatic tricks for anxiety'. Disabled, queer and trans people have been at the helm of much of the conversation on radical, justice-centred care work, often catalysing change through writing, storytelling, zines, online calls and conversations around dinner tables.[10] They know what it means to disrupt and make change within the bounds of their physical, emotional and mental well-being and capacity. They know what it means to honour themselves and each other, to be vulnerable and compassionate yet deeply rebellious.

We need a care revolution, not only within our economic and social systems but in our movements too. As Leah makes clear, we must acknowledge the way in which 'all bodies are unique and essential, that all bodies have strengths and needs that must be met [and] that we are powerful not despite of the complexities of our bodies but because of them' and that we must 'shift our ideas of access and care (disability, child care, elderly care, economic care) from an individual chore, an unfortunate cost of having an unfortunate body, to a collective responsibility that is maybe even deeply joyful'. Without engaging with care in this way, we make invisible and undervalue the real care work that grounds our world day in and day out. The care work that continues, through the most impossible conditions, to keep us all safe.

31 Climate Work Is Care Work

Policymakers, politicians, scientists, lawyers and billionaires. These are the people we believe our future is exclusively governed by. Many of the individuals in these roles sit at the centre of decision-making rooms, theatres or halls. They negotiate, barter and lobby. Pushing for more stringent action or pleading for watered down messaging. On the other side of the door, banging and hacking at the woodwork, are the activists demanding more, better and faster. And herein lies our most common image of climate work, the constant struggle between the rich and powerful, the bureaucrat and the activist. Aside from the overwhelming presence of fossil fuel execs, these are the people who predominantly make up the attendee list at events like the COP conferences. I started writing this book in the wake of COP26 in 2021. Being there in Glasgow, it was obvious that the formal negotiations, despite the incredibly hard work of the negotiators, were failing and that this failure was most directly linked to the proximity of politicians, multinational corporations and, most frustratingly, fossil fuel companies. When a global climate conference sees more fossil fuel delegates in the formal negotiating arena than global Indigenous delegates, the priority of that conference becomes very clear. A priority that continues to be made clear – as I write, the COP28 conference has welcomed nearly 2,500 fossil fuel delegates, nearly four times as many as the previous year.[1] The blatant neglect of care and commitment to people and the planet was painful. The consensus from climate activists and commentators was that the meaningful action taking place at COP26 was not happening inside the Blue Zone, the area that hosts the formal conference negotiations and where the most important action *should* have been happening, but was taking place outside, on the streets. By all accounts, COP26 was a bubble. One with thousands of climate activists, scientists, policymakers and creatives all vying to crown themselves as the 'real' leaders, or as having the only 'real' impact. The inner politics and self-aggrandisation felt at odds with the possibility of being supported by a global gathering of passionate and dedicated people. Instead, what I saw were tired, exhausted, burned-out faces battling nonchalance and obligation with fierce desperation. Most activists I have spoken to have agreed that these spaces – whilst essential, especially for Global South change-makers who rarely have a chance to be heard on a global platform – are places void of care. There for only five days of the two weeks of the conference, I was already becoming a shell of myself, and I couldn't help but think of all those who couldn't even make it to the conference. Those hindered by the extortionate cost of travel and visas, those blocked by discriminatory immigration rules, those with disabilities that prevented their travel, those who had duties of personal care they could not get away from. I thought about the work I wanted, needed, to get back to, and the work of so many others that had continued, invisible, during the conference's proceedings. I wondered what the conference might have looked like if care was centred. If the legacy, experience and strategies of nurturers and care workers were centred and amplified, guiding the way we think and act in decision-making spheres. It was no surprise that this was not the case. Even though there were mothers, disabled folks, healthcare professionals and other care workers in attendance, their numbers were dwarfed. Care was rarely centred in conversations and when it was, it was often Indigenous delegates impressing the importance of sociological and ecological nurturing as a central climate solution. Intentionally or not, the noise of the negotiations and marches alike – in all their necessity and value – drowned out the small whispers calling us to more tender, attentive and compassionate ways of being. Unlike the high visibility of mainstream movements and politics, and grandeur of large conferences, most care activities go undocumented, unseen and unappreciated.[2]

The climate crisis is a threat multiplier, meaning that its impacts exacerbate the suffering experienced by vulnerable and marginalised communities – women, children, the elderly, disabled folks, poor folks and small-scale land workers in the Global South. The same communities suffering at the hands of rampant consumerism, individualism and a dysfunctional care system also suffer the worst effects of both the care and climate crises. Many of whom undertake the majority of our societal care load, whether by nurturing the next generation or growing the world's food. It is those who are dependent and depended on that have an intimate understanding of the necessity and precarity of care. They feed and nourish the world whilst going hungry. They are the cherished 'future generation' facing an ever more hostile and unsafe world. They are the first on-the-ground responders, yet the last to be rescued in the wake of environmental disasters.

In 2017 a deadly October wildfire storm saw fourteen large fires sweep across the state of California, including the Atlas Peak fire that rushed across the Napa Valley consuming everything, including houses, in its path. In one of those houses, two people lay in the path of death's scorching fingers. Teresa Santos was the home-care provider for the immobile ninety-year-old Sally Lewis. Sally's immobility and the lack of suitable rescue services in the area meant that her fate was already sealed, and Teresa stayed by her side, refusing to leave her patient to die alone.[3] Teresa and Sally's tragic story is emblematic of the real dangers faced and losses felt by care workers and vulnerable people in the face of environmental disaster. Poor, elderly, disabled folk and those living within institutions – along with their carers – are often the last to be evacuated during environmental disasters, seen to be 'expected losses'.[4] It's the care workers – the doctors, migrant farmers, nurses, therapists, teachers, cleaners, firefighters and paramedics, often women, immigrants and people of colour – who become first responders on the ground even while they experience the impacts of environmental disasters and become victims themselves. Caretaking jobs continue to provide environmental and social value across the globe; they just remain grossly under-appreciated, underfunded or undermined. In times of disaster, care workers become connectors and pollinators, essential sources of knowledge for their communities. As Kim Evon, executive vice president of the Service Employees International Union Local which represents over 400,000 care workers, explains: Care workers 'become the connection point for fire and police during wildfires and [power] outages because they know the conditions of people who are disabled, who are immobile'.[5] Their work at this time can be deadly. We saw how this plays out during the Covid-19 pandemic: Underfunded healthcare workers undertook the immense job of keeping alive as many people as possible, while working through gruelling conditions that inflicted extreme mental and physical strain as well as posed the very real risk of death. The threats faced by care workers during the pandemic came not only in hospitals or on public transport, but also in fields and farms across the world. In 2020, California saw wildfire hell, with 4.1 million acres of land set aflame by 10,000 individual fire incidents.[6] At the time of writing, it was considered as the state's worst fire season ever. While many were confined to their homes and thousands were evacuated, migrant farmworkers faced the unthinkable. An apocalyptic, hazy sky filled with orange clouds above them and ash-laden soil and crops beneath them, thousands of people, found to be disproportionality impacted by coronavirus, were forced to continue nourishing the world at the expense of their own safety, classified as essential workers. They worked, without the provision of masks, with ash and dust whirling around their bodies, settling into their necks and making it impossible to breathe. But there was little choice – advocate for your safety and risk losing your job. No protection from the ash, from the smoke, from the heat or from the heavy pesticides they are forced to work with. The reward? $5.50 for an hour's work.[7] For these farmworkers, the violation of their health and safety extended beyond the pandemic, it is practically business-as-usual in the industrial agricultural industry. Speaking with the *Guardian*, thirty-eight-year-old Reyes, a strawberry-picker, explained that 'there is also a culture where if you speak up or say you don't want to work, you may be seen as someone who is lazy or doesn't want to work and you may not be called back for the next harvest'. Land work, such as strawberry picking, is incredibly labour intensive, and it was on farms of demanding crops where the most suffering was seen, with hundreds working ten-hour days with no protection.

In the UK, investigative research from Wicked Leeks, a publication from the organic farm Riverford, and the media platform Live Frankly, exposed the horrendous working experiences of migrant workers on British farms. One of the three migrant workers interviewed by an investigative journalist shared that 'We weren't humans, we were chattels', referred to as their assigned numbers rather than their names.[8] This is the account of just one of the 70,000 seasonal workers that help feed Britain every year; incidentally, since the Brexit policy came into place, the UK has struggled to find as many workers as needed.[9]

Landworkers, farmworkers and growers are carers, providing sustenance for society and for the soil; yet, the capital-obsessed, uncaring industrial farming industry has traded nurture for profit at the expense of the well-beings of humans and the environment. It exploits and exhausts land-based care workers and the land in equal measure. The global industrial food system is responsible for 70 per cent of the water extracted from nature, the cause of 60 per cent of biodiversity loss and can generate up to a third of human greenhouse gas emissions.[10] Despite the promises of the Green Revolution, which took place between the 1960s and 1990s when technological advancements and selective breeding led to increased crop yields, the gains made during that time have been met with significant losses.[11] Our increased 'productivity' has rightly brought millions out of hunger but at the same time has fuelled a hunger crisis by depleting soil health and contributing to climate change. This is already decreasing yields, affecting most acutely subsistence farmers in the Global South and often women. Almost 40 per cent of the global population is dependent on agriculture for their livelihood and sustenance, with these same people producing over 50 per cent of the world's food.[12] In lower-income countries, women make up 43 per cent of the agricultural labour force.[13] In some regions such as South Asia and sub-Saharan Africa, women are the majority of the agricultural workforce. As the climate crisis worsens and the planet experiences increasingly extreme weather events like floods, droughts and forest fires, it is these women who are severely impacted, having to travel further distances, toil the land harder and face increasing food scarcity. These women are often balancing care duties, looking after the land, their children and their communities. But these women have also had enough: They are fighting back and re-rooting care and nurturing back into the land practices.

Across twenty-two states in India, women have come together to grow 'real living food', not as a commodity but as a source of life. They are members of Navdanya, an 'Earth-centric, Women-centric and Farmer-led movement for the protection of Biological and Cultural diversity', founded by Vandana Shiva, a pioneering eco-feminist, scholar and activist. The organisation has built a 150-community strong seedbank, reclaiming their traditional foods and plant biodiversity destroyed by large pharmaceutical companies who practise biopiracy, the process of appropriating Indigenous biodiversity knowledge systems, without permission or compensation, for profit. These corporations can apply for and be granted patents, characteristically awarded for 'novel' inventions, for seeds or biological knowledge that they sourced from Indigenous communities, knowledge held by those communities for millennia. The result is that monopolies are created to control resources Indigenous communities previously used to meet their health, nutrition and cultural needs, requiring them to pay for what is and has always been freely available to them. The consequences have been dire: As many as 250,000 farmers have died by suicide in the last decade over failed cotton harvests they had taken out loans for, to afford patented cotton seeds. Navdanya, meaning 'nine seeds', works to heal communities and lands across India by sowing sovereignty through their seed bank as well as agro-ecological and Indigenous farming methods.[14] They are reclaiming and regenerating a food system that honours the act of growing, nurturing and nourishing the Earth and its people. A system that recognises land work as care work, as so many Indigenous communities around the world know and do.

As the climate crisis worsens, investing in strong, resilient social and environmental care systems will become crucial to our collective survival and flourishing. Rapid and radical investment in our care infrastructures will only become more important as we move towards an economy that is decoupled from carbon. Our transition away from fossil fuel extraction must happen quickly but equitably. With an estimated 73 million jobs set to be lost from high-polluting industries, a significant task over the next couple of decades will lie in creating a foundation of well-paid, fulfilling, low-carbon jobs.[15] Around 46 million of the jobs lost in the fossil fuel sector will be replaced by roles in the booming renewable energy industry. Additionally, a tripling of investment in jobs centred on the restoration of life and biodiversity in forests and oceans, nature-based solutions, could create as many as 20 million jobs by 2030.[16] Achieving a just transition will require investment and innovation across all sectors, but governments and corporations continue to proceed blinkered, focused only on 'hard' climate infrastructure. Think wind turbines, electric cars and charging points, low-carbon shipping or retrofitting housing stock. Reducing emissions through these approaches is essential, especially given their overwhelming contribution to global carbon emissions. But as we have come to know, the climate crisis is not only a crisis of carbon. As those in power argue, stall and poorly execute 'hard' climate infrastructure, pouring billions into necessary and essential energy grids, train lines and flood defences, our collective infrastructures of care, which are equally essential to climate resilience, remain neglected. As US congressman Jamaal Bowman makes clear, we must 'treat care work as what it is – infrastructure – and invest in those jobs and workers so that we can build a truly sustainable economy'.[17] No longer can care work be taken for granted. Cast to the side-lines, relying on the passion and sacrifice of caring individuals whilst exhausting the Earth and its inhabitants. We forget that jobs won't only be lost as a result of transitioning to renewable energy but through the impacts of climate and environmental breakdown too. The International Labour Organisation estimate that in 2030, the equivalent of 80 million jobs will be lost due to heat stress alone.[18] Faced with climate and environmental breakdown, we have an opportunity to rethink what good infrastructure looks like and to expand the definition and creation of 'green' jobs. The advantage of investing in a caretaking economy is that it already exists. We not only need new engineers, technologists and construction workers but also well-compensated, healthy and fulfilled doctors, nurses, emergency workers, psychologists, and child and elderly care workers (including mothers, parents and guardians) to provide the essential care that will increasingly be needed in a world impacted by climate chaos. Investing in these, and the many hundreds of other unmentioned caretaking roles, is also an essential part of delivering a just transition. There will be millions of individuals, mostly men, who will be unable to access the new stream of clean energy jobs and be left searching for alternative work and meaningful new careers.[19] Necessary work will need to be done in order to move past the social myths that label care work as women's work, as undignified, unrewarding work. Myths that harm the men that *already* work in all types of care role. Instead, care jobs must become well compensated, as described by the team at the Feminist Green New Deal in the report 'Care & Climate: Understanding the Policy Intersections'. Written by Lenore Palladino and Rhiana Gunn-Wright, the report highlights the necessity of uprooting the patriarchal belief system, one with deep roots in the slave economy, that devalues care work and makes it unattractive or unimportant for the men in our society. Uprooting these belief systems requires us to make care work fulfilling, with care workers provided the right to unionise, the ability to comfortably support their families and the space to use and develop the skills they bring from other industries. These investments are not only good for society but also provide huge economic benefit. In their report, they highlight a recent analysis conducted by Palladino and their colleague Rakeen Mabud which found that a $77.5 billion annual investment in care would support over 2 million new jobs, translating into $220 billion in new economic activity, even before job quality improvements are factored.[20] As the climate and environmental crises worsen, we will need more and more people working in and strengthening our care infrastructure. People who are themselves healthy, provided for and given the resources to do their jobs safely and with dignity.

If we are to survive the end of this world and successfully usher in the beginning of the next – emerging healthier, remaining rooted and experiencing more abundance – we must acknowledge that resilient communities are born from robust, strong and thriving social infrastructures. What becomes clear is that safer, more just futures don't just depend on electrification or carbon sequestration, but a radical reimagination of how we structure our society and the way in which we engage with, support and invest in care activities. As the care scholar Joan Tronto makes clear, care is 'a species activity that includes everything that we do to maintain, continue, and repair our "world" so that we can live in it as well as possible'.[21] While many of us have grown up to believe that the solutions to suffering and disaster must come from the outside, from 'someone out there', or from those in power, for most marginalised communities, the radical change they need can only be provided from building resilience and solidarity together.[22] And so, care work, essential climate work, must transcend profession or biology and become our collective, continuing, unceasing work. To do this, we must, as Katherine May told us, begin to blur the lines between ourselves, each other and the Earth. We must become kin.

32 Making Kin with the Earth

On the morning of 29 August 2005, the world ended. If you recognise the date, you already know how. The sky darkened, painting every person and surface in a wash of blue-grey. It swirled, and spat, and ripped, and sprayed. It summoned the wind, tormented and teased it into such a fury that blown-bare trees seemed to flicker like objects caught in the matrix or like the mirages that trick us in the sweltering heat. And then the water came. Gushing, surging, bursting, breaking and breaching levees, racing into the city. Rushing to fill every crack and crevice, every ditch and every corner of every home. Once the water could no longer race outwards, it rose upwards. Chasing people as they climbed to safety, nipping at their heels, bullying them, pushing them higher. It wasn't satisfied until the last gasps of the city were silenced, confined to a life under water. The roofs became rafts and it was there that people stayed. Waiting, hoping, praying that someone would come and rescue them. 29 August 2005 is the day that Katrina came. A day I do not remember clearly, as most other seven-year-olds would not. But a day I have been transported to by the articles, films, documentaries, survivor accounts and books that document the tragedy of that day.

When we picture the end of the world, some of us imagine that the police will be the first to come to our rescue. We pray that the emergency services might have enough capacity to save us. Failing that, we'd wait for the military. Some of us have lost faith in humanity, bypassing the option of being saved, preparing for the inevitability of violence and fierce competition, steeling ourselves for the brutality we believe will await us. Yet, in the poorest, Blackest neighbourhoods of New Orleans, what prevailed in the wake of Katrina was a system of care propped up by those left most vulnerable and invisible during the disaster. After two days without power, food or rescue, stranded in his home in the Fischer public housing complex, a barely adult Jabbar Gibson emerged from his flood-threatened shelter in search of gas and a vehicle. Discovering a dozen school buses in an open depot, and finding the office unlocked, Gibson and his neighbours raced to find a bus that was still working. Bingo! One of them had been left with a full tank. Gibson returned to Fischer, eager to pack as many of his sixty family, friends and neighbours in, many of whom were sick, elderly, disabled or pregnant.[1] In an unlikely turn of events, the police arrived, not to take the residents to safety but to reprimand them for theft of public property. An argument ensued, not least because Gibson was a known criminal with multiple drug offences. But when Gibson's mother pointed out that it wasn't the police who were helping the vulnerable but her son, they handed them the keys and went on their way. After twelve hours of evading police blockades, decimated roads and carrying the precious lives of sixty others in his hands, Gibson, who had never driven a bus before, delivered everyone safely to Houston's Astrodome, where they finally received food, water and medical care from the tireless Red Cross volunteers.

Gibson's is only one of many similar stories of bravery, courage, community and care that emerged in the wake of disaster. Across New Orleans, communities rich and poor were coming together to care for each other where the state could not. Out of the debris rose one such community, Common Ground, who came together to honour the calls of Malik Rahim, a former Black Panther. Growing out of Rahim's kitchen, the group became the glue that held the people of New Orleans' lower ninth ward together.[2] Upholding proudly the legacy of the Black Panthers, they fed large groups of people, set up independent media centres, provided street medical aid and used their collective skills to pick up the pieces, and each other, after disaster had struck.[3] The story of Common Ground, like that of Gibson, is not straightforward – the care capacity of the organisation unravelled as its internal links and connections deteriorated. But both of these stories, in all of their complexity, are important in uncovering the roots of collective care. These stories tell not of a saintly care. This is not the kind of care that is neat, clean or polite. It's the kind of care that is gritty, ragged, complex, unnerved, unhinged and essential. What is buried in the chaos of Gibson's story, a story of a convicted drug dealer and his vulnerable community, and the story of Common Ground, in its intense impact and tumultuous unravelling, is the reality of care that those who have grown up working class know all too well. That care is given and received regardless of perceived worth, admiration, behaviour or popularity. Care is unconditional. Care moves, as the Care Collective writes, 'across difference'.[4] Katherine May and I reflected together on our own experience of this, not the tragedy of environmental disaster but the indifferent, unflinching care we experienced and observed on our respective council estates. We laughed at the humour that often comes with this kind of care and also the radical honesty. The begrudging grunts and stifled yawns that say, 'The bloody cheek of knocking on so early, but yeah chuck the kids in the living room I'll get them to school', or the unjudgemental, undocumented feeding of other people's children, with little more in return than thankful eyes and the knowledge that the same kindness will be unquestioningly reciprocated when you are next in need. Buried in the madness and dysfunction of a council estate lies part of the blueprint for care practices needed to build resilient communities. As Leah Lakshmi Piepzna-Samarasinha writes, 'Community is not a magic utopia, just like our families aren't, but we don't need to rely on being liked or loved or happy or amenable to [give or] receive care.'[5] When disaster strikes, our entanglements are laid bare. We are temporarily stripped of the labels our conditions or society force upon us, but we become more intensely human, knowing that our survival is inarguably linked to the survival of our neighbour.

(Un)natural disasters – environmental, social and political – often mark us with wounds and scars. They cover the land, our communities and our spirits with cracks and through them we come to truly see each other. The veil of individualism is lifted, even if for a moment. But how do we translate these ephemeral moments of interdependence into a way of being? In the Western worldview, we perceive care as an act that can and must end with our recipient becoming better, being cured and no longer needing our support. We become suspicious or fearful of care's ongoing nature, the after-effects of suffering and the necessary continuity of need or mutual aid. As described by Leah Piepzna-Samarasinha, able-bodied Westerners have become accustomed to a 'crash-and-burn emergency model'. When disaster strikes, someone falls seriously ill or is hurt, there is a huge wave of urgency. We overcommit ourselves to action in order to find purpose. Our zeal for service, for expending unconditional energy on our loved one, has a shelf life of a few weeks or months and then rapidly trickles off. We expect the hurt and suffering to be gone or are just too exhausted from the overpouring of care and attention that we hoped would be powerful enough to *fix* the situation. As Piepzna-Samarasinha writes, 'These emergency-response care webs often really fall apart when and if the person they're caring for becomes disabled in a long-term way... when they realise that the "issue" isn't an individual problem' but something that becomes shaped by the inaccessibility and hostility of society.[6] We must come to understand that disaster is not only that which shocks and startles us. It is not only that which rocks and shakes the world in one dramatic blow. For many people, disaster happens every day. Disaster is the blood-curdling, irritating beeping of an empty prepayment electricity meter. It's the teasing and torturing orange flashing of the 'check engine' sign on the car dashboard. It's the heaviness of hunger when having to pay for warmth at the expense of food. As the Fiji-based climate organiser Thelma Young-Lutunatabua described in a conversation with the writer Rebecca Solnit, for those who experience ongoing suffering, often the only way through is together; held by the practice of mutual aid.[7] Mutual aid is a long-term, grassroots practice of building community resilience in the face of disaster and ongoing crisis. It is a process of survival, with webs of care stepping in when systems fail to meet the needs of its people. Communities around the world, across north–south, east–west divides, have a history of practising mutual aid – throughout history it has been less of a practice and more of a way of life – but we can trace its formalisation to the late 1700s when mutual aid societies were established by the Free African Society as a response to the abandonment of newly freed Black Americans by the US government.[8] It was later introduced and popularised in Western social literature by the Russian anarchist and geographer Peter Kropotkin, who advocated for an alternate story of human nature. One which acknowledged our acts of hatred, violence and competition but also gave space to celebrate, appreciate and become curious of our equally unceasing practices of interconnected care. As the scholar and film producer carla bergman makes clear, mutual aid is not something we need to learn about or create from scratch, 'It has always been a part of our collective and ancestral memories.'[9] Our task is to recover, remember and enact collective care as an ongoing practice, weaving webs of nurturing across our communities beyond charity and towards solidarity.

Mutual aid is a practice familiar and central to poor, Black, Indigenous, disabled and queer communities. Communities with either very little access to state care or for whom the need for care is inseparable from the trauma of death, isolation or disownment at the hands of uncaring people and systems. In the queer community, mutual aid flowed from the bosom of the ballroom, which grew out of the Antebellum South where enslaved people would pantomime their masters at dances and became a safe space 'of queer kinship, performance and care'.[10] Often headed by trans women of colour, who acted as other-mothers for unhoused and vulnerable queer people, care spanned from providing food and friendship to free legal services and grief counselling, especially through the HIV/AIDS epidemic. In the African American community, one of the most famous and successful practices of mutual aid came from the Black Panthers' Breakfast Programme. Beginning in 1969 at an Episcopal church in Oakland, the Panthers would grocery shop with the direction of nutritionists and prepare and serve free breakfasts before school for the local children. From this one programme sprouted forty-five, with thousands of children being fed every morning, and thousands of families receiving access to medical and legal clinics and community ambulance services.[11] The impact was huge, with school principals reporting that the children were more attentive and less exhausted and distractable in class. It also created space for community organisation, for education on the Black struggle and was a way to organise people for activism. So threatening was this relatively simple act of care that the FBI, led by J. Edgar Hoover at the time, called it 'potentially the greatest threat to efforts by authorities to neutralize the BPP and destroy what it stands for'.[12] By caring for their community, the Black Panthers were fuelling 'revolution by encouraging black people's survival'.[13] The act of caring for one another, whether in the ballrooms of New York, the churches of Oakland, the wards of New Orleans or the council estates of London, is and has always been, an act of rebellion. An act of revolutionary love that transcends the nuclear family, that transcends genealogy. It is a love – in all its gnarled and knotted complexity – that is transported steadily, unceasingly and unceremoniously through the roots of a community, if only they have been well tended. It's a love that urges us, as the feminist scholar Donna Haraway beautifully writes in her book *Staying with the Trouble*, to learn how to live and die well with each other.

In her book, Haraway asks us to *disturb* and *stir up trouble* in our societal ideas of what it means to be *kin*. To rebel against the West's cult of individualism and suspicion that tells us there is 'no sense in having any active trust in each other, in working and playing for a resurgent world'.[14] She asks what it means to *make kin*, to be *kind*, *to become kindred*, to 'have an enduring mutual, obligatory, non-optional, you-can't-just-cast-that-away-when-it-gets-inconvenient, enduring relatedness that carries consequences'.[15] For me, the question of making kin is central to our ability to 'survive' ecological collapse, emerging as unscathed as possible and, ideally, more connected. How do we come to live in a state of response-ability – a practice introduced by feminist scholars such as Donna Haraway, Natasha Myers and Karen Barad – that goes beyond the obligation of responsibility and moves towards being *responsive*, in an empathetic and emergent way, with the rest of the world? How do we respond to, with and for all our earthly relatives? How do we become more attuned to the inescapable, invisible threads that bind us in an infinite tapestry with all beings on this planet? To see the bird in the banker, the life in the stone, the microbe in the madness, the animal in the human.[16]

Aanikoobijigan

| *Great-grandparent/ ancestor.* | *Great-grandchild/ descendant.* |

The word that the Anishinaabekwe-Ukrainian writer Patty Krawec and her community use to describe a person three generations before them also describes a person three generations after. As she writes in her book *Becoming Kin*, the term *Aanikoobijigan* connects seven generations.[17] Krawec makes clear that 'we are in the midst of beings with whom we are in relationship, whether we acknowledge it or not'. She also describes how, depending on the prefix or suffix used, the word also describes the act of sewing or the tying together of things. To become kin is to stitch ourselves not only to all beings on this planet but all those who have come before and who are yet to come. How do we claim our ancestors and how do we claim our descendants? Furthermore, how do we acknowledge the land as a relative, past, present and future? To quote the Kānaka Maoli (the traditional name for native Hawai'ians) genomic researcher Dr Keolu Fox: 'The land is my ancestor, that is a scientific statement.'[18]

In his paper 'Kincentric Ecology: Indigenous Perceptions of the Human–Nature Relationship', the Rarámuri scholar Enrique Salmón reflects on his homeland Gawi Wachi – the Place of Nurturing – in Mexico, one of the most biologically diverse regions in the world. He reflects on the meaning of *iwígara*, a concept that, like the landscape the community lives in, is multifaceted and diverse in meaning but that encompasses the community's 'interconnectedness and integration with all of the life in the Sierra Madres, spiritually and physically'.[19] *Iwígara is* accompanied by *numatí*, which holds that all things of the natural world are relatives. That the 'peyote, datura, maize, morning glory, brazilwood, coyotes, crows, bears, and deer' are as much the Rarámuri's relatives as their bloodkin. Not in a fantastical or proverbial sense, but with a genuine acknowledgement of this fact as reality. What emerges from feeling such deep connection to the living world is not awe or wonder but as Salmón makes clear, a profound familiarity with the living world. That then must become our collective objective, to build a radical intimacy with the soil, with other animals and with each other – with our *kin*.

In her essay 'When You Could Hear the Trees' for *Emergence Magazine*, the Northern Irish writer Kerri ní Dochartaigh gives us a glimpse of the way in which motherhood and the experience of pregnancy, of growing and nurturing life, bring to the fore questions regarding the animal within us and our relationship to the land. She describes how pregnancy 'unsettled' her understanding of what it meant to be human. Her transition from maiden to mother revealed something 'ancient . . . feral . . . mammal'. It was in trying to give voice to motherhood that she was in fact 'trying to give voice to mammalhood', experiencing like never before being a part of an ecosystem. She highlights the rebellion of listening to herself and the Earth through her body, disregarding our collective Western agreement to find no similarity, only superiority, to the wild, the animal, the creature. Over a video call, Lucy Jones, author of *Matrescence*, a radical exploration of motherhood, shared her own experience of birthing and mothering with me, explaining how she found her experience to 'profoundly alter [her] sense of being rooted to the world', in a way that went against her very rational and engrained post-Enlightenment way of thinking. She reflected on her childhood, finding that despite her family loving the natural world, it was a love that was based on and influenced by patriarchal ideas of domination and intellectual supremacy. Of humans having 'grown out of our connection from the Earth'. But after a forty-six-hour labour, she felt faced with the reality that she was 'something wild and uncontrollable, animal'. Something she had been running away from throughout her maidenhood. Birth and mothering, of course, are not the only ways to blur the boundaries between our bodies, our minds and the Earth. We must, as the writer Jessica J. Lee shared with me over the phone, move away from the lazy, normative conversations that see care and nurturing as the work of mothers alone and that at the same time vilify and exclude the unique learnings motherhood brings to the way we conceive our place in relation to other beings. What the experiences of not only birth or motherhood but the nurturing of children bring us closer to is the way in which they confuse our definition of what it means to be human. That, as Lucy described, the experience of looking both life and potential death in the eye interrogates and dissolves the margins we desperately defend between ourselves and the world. That these experiences remind us that we are of the soil and that we will return to it.

Whether through parenthood or not, we must all find the portals that transport us out of our self-centredness and supremacy towards symbiosis with each other and the rest of the living world. In fact, one of these portals is never far from reach, accessible to us all. Compost. A substance that simultaneously embodies death and creation. Decay and abundance. Loss and life. To many, compost is just dirt. But it is also a planet, an organism, an animal. It's a life-bringer and a killer. Within it lie the lives of hundreds if not thousands of microbes, critters and crawlers; yet, it is also the refuge of seeds and sprouts and forests. Forests that out of thin air, our air, rise from the soil. Forests that feed us and fill our lungs. How then do we separate ourselves from these beings? If our breath makes the forest and the forest makes us and the soil makes the forest,

are we not the forest,
                    and the soil
                                    and the breath?

We can start this train of thought from an infinite number of objects or beings. We can take a car and arrive at the lithium, the steel, the aluminium – at the soil. We can take a cup of coffee and arrive at the bean, at the plant – at the soil. As we travel down these myriad paths, returning again and again to the soil, we engage in the process of *making kin*. Haraway tells us that to make kin is to make-with, become-with and compose-with. And as we do so, our edges become blurred. Where we begin and our kin start becomes unclear. Lucy Jones shared with me her love for and growing obsession with slime moulds. She described how they start out as a 'Plasmodium' – a slimy sheet, a 'slick matter' of cells that is sometimes colourful and at other times colourless – that, animal-like, creeps and 'scoffs and predates' on the ground.[20] She describes how this Plasmodium is capable of learning, anticipating and teaching, transforming, if it wants, into something completely different. They become mushroom-like, 'extremely beautiful and intricate and iridescent'. By making kin, like slime moulds, we become shapeshifters and tricksters as Báyò Akómoláfé told me, becoming 'stewards of queer hope, coaxing ourselves away from safe grounds to the monstrous ambiguities of being more fully present'. By making kin with the Earth, we become everything and nothing at the same time.

Epilogue

ROOTED HOPE: Our Natural Connection

The rain came slowly at first. Tap.
Tap

Tap.
Tap.
Tap.

Tap. Tap. Tap. Tap. Tap. Tap. Tap. Tap. Tap. Tap. Tap. Tap. Tap. Tap. Tap. Tap. Tap.
Tap. Tap. Tap. Tap. Tap. Tap. Tap. Tap. Tap. Tap.
TAP. TAP. TAP. TAP. TAP. TAP.

And then it came all at once. My vision through the windshield of my uncle's car, at first obscured by the clouds of red dust thrown up by the vehicle ahead of us, was now distorted by a bubble-wrap pattern of quick and heavy raindrops, light-bending rings of water on the glass. The sky was falling.

We were headed to my paternal grandparents' home in Accra, Ghana, which I had not visited since I was a child. The road, which just moments earlier had been dry and dusty, was now being assaulted by the heavens. To my right lay a steep incline packed with houses, shops and abandoned buildings; from each crevice, opening a path around them, gushed a rush of murky brown rivulets, racing to the land below. We made it into my grandparents' compound just in time. The water crept, snakelike, beneath the main gate and chased us to the front door. We retreated to the safety of the main living room, pulling back the netted curtains and our eyes locked upon an astonishing scene. The house had become an island. All around it stood a lake of water threatening to engulf the raised veranda.

We were caught in a flood.

It wasn't until six hours later that the water had receded enough to lay makeshift stepping stones, which rose just above the surface of the water, and climb our way over the gate to the other side. Up until that point, I had been privileged enough to have never personally experienced a flood. I had watched in horror as images and videos rushed in from the extraordinarily destructive 2015 floods in Accra that affected over 46,000, displaced over 9,000 and killed 200 people as well as intensified the cholera epidemic.[1] As the climate crisis worsens, the floods become more powerful, more destructive. The situation is dire, and Professor Christopher Gordon, a founding director of the Institute for Environmental and Sanitation Studies at the University of Ghana, tells us that without radical action 'the worst-case scenario is that practically half of Accra will be unliveable—all will be damaged'.[2] Ghana's resilience is weakened by poor land management, with swathes of forest and wetland, ecosystems that play an essential role in trapping, storing and absorbing floodwater, being converted into real estate. Root systems allow water to penetrate deeper into the ground around trees, providing the planet with natural flood management.[3] But Ghana, like many countries around the world, is losing its roots. As the trees disappear, the floods worsen and our ability to survive becomes threatened.

Nowhere is this problem more prevalent than in southwestern Bangladesh, a region that is regularly destroyed by tropical storm surges, some of which have reached heights of 3 metres. The year 2021 saw the region being battered by Cyclone Yaas, which affected 1.6 million people and damaged 26,000 homes.[4] In the past, the region's shoreline had been protected by the Sundarbans, the largest mangrove forest in the world and a powerful protector of the coast.[5] The mangroves act as natural barriers that not only stabilise soil sediments with their roots but also buffer the impact of tidal surges.[6] But with the worsening climate crisis, stronger and more frequent cyclones are decimating the mangrove roots. The storms are also accelerating land erosion, stripping the plants away from their home; a quarter of the mangroves have already been lost.[7] The situation is urgent and efforts to restore and regenerate mangrove systems is intensifying.

Floods tear apart the land and our lives. They cause everything to come undone and leave us grieving, with no choice but to remake our worlds. In her powerful book *Becoming Kin*, Patty Krawec reflects on the persistence of flood stories, alongside creation stories, in cultures around the world. She tells us that while creation stories tell us how we began, flood stories teach us how to rebuild. She tells the story of Nanaboozhoo, the shape-shifting trickster and creator who one day awakens and finds himself floating in the water on a log with a group of animals. He realises there has been a flood. Water surrounds him as far as the eye could see. He comes up with a plan. Remembering the Anishinaabe creation story, which tells of the first woman who fell from the skyworld and with a handful of mud created the Earth, he dives into the water and attempts to reach the ground. He has no luck, and gasping for air turns to the animals. A small muskrat volunteers and disappears for a long time. At last, he arises through a cascade of bubbles, no longer breathing but clenching in his paws 'the mud that became the land ... and the world was made anew'.[8] The future lies in the mud, in the soil where the roots rest. The story shared by Krawec, a story of collaboration and redemption, is also a story of grief. Grief that is ever present in our own collective story. Grief, she tells us, is 'the sound of thunder you feel deep in your chest' and also 'the persistence of love'.[9] This grief, a grief that holds within it both despair and hope, must take us deep into the flood. Far enough so that *we* might also grasp a handful of Earth, and surface with the makings of a new world. We have, to various degrees awoken as Nanaboozhoo did, to find ourselves drifting on floodwater. With future deluges making their way over the horizon. This awakening has been a slow and confronting experience. We teeter on our rafts, unsure whether to make the leap. Silencing the rumbling, crashing and thundering in our hearts, many of us find comfort in admitting defeat.

Why must we dive when we are surely going to die?

We stubbornly continue life on the raft, waiting to be submerged by the rising tide. As Krawec writes of our survival, we tremble with fear asking, *What if we can't swim?* Without asking,

What if we can?

Paraphrasing Mark Fisher, the author of *Capitalist Realism* who gave us the famous line 'It is easier to imagine the end of the world than the end of capitalism', it seems that many find it is easier to accept the end of the world than to hope

and act for its survival.[10] While conversing with Rebecca Solnit, we reflected on the nature of hope and the irony of a specifically Western, middle-class hopelessness. Overwhelmed by the anticipatory grief of climate chaos, not yet having been physically or personally impacted, those living relatively safe lives give up all hope of a liveable future. What Rebecca finds striking is the way in which 'people on the frontlines don't surrender easily, because surrender for them means your children are going to starve to death, you're going to live and die in a gulag, or a displaced persons camp, you're going to lose everything, and real horrors are going to happen to you'. While those experiencing climate disaster first-hand fight for their lives, and in so doing fight for ours too, we indulge in the biased, destructive and deeply negative news cycles or well-meaning but scaremongering social media videos, all of which etch a permanent image of hopelessness and despair in us. We lap these images up, feed on them and spew them back out into the world. They fester, pollute, foam and froth at the edges of our mouths and consume our minds.

As we talk, Rebecca asks me to imagine that we are in a sinking ship, thrashing and panicking in the water with countless others. We see some lifeboats in the distance, yet many around us are certain we are all going to drown, that there is no sense in swimming to the lifeboats. But, if there are a hundred people in the water, if we get to a lifeboat, maybe an individual can pull six kids out of the water and another can rescue another three: 'It's not a total victory, but it sure is victory for those nine kids.' When we are faced with the possibility of hope, we discard it as naïve, frivolous and out of touch. We point to the world's many terrors, its ugliness and pain. Not realising that in our defence of the end of the world, shines through our deep care and grief for that world. Care and grief we have thrown deep into the water, to be buried far out of sight, too fearful of the unknown potential they hold. Certain failure feels much more comfortable than uncertain triumph, than to act without the knowledge or promise of safety. Rebecca tells me that 'defeatism is a form of certainty and hope is a form of uncertainty, hope is that we can get into the lifeboat, and it leads to possible survival or possibly being wrong'. Defeat tells us that if we can't save everything or everyone then we can't save anything. Although despair is a very natural response to the immensity of the interlocking social and environmental disasters we face, Rebecca reminds us that it is 'best not confused with an analysis'; you can 'feel terrible and remain committed, be heartbroken and know the future is being made in the present'. We often confuse hope with optimism, the idea that everything will be okay, but optimism and defeatism hold the same space. Both outlooks, she says, 'are a form of confidence in that you know what is going to happen and therefore nothing is required of you'. Her comments reminded me of a conversation I had had with Báyò Akómoláfé, who highlighted the perils of divorcing hope from hopelessness, of the incapacity to hold them both simultaneously. He described how blind hope 'pays no mind to shadows, dips and grooves, places we don't know' while 'hopelessness released from hope starts to misbehave. It becomes this navel-gazing pessimism that is all doom and gloom, that also does not notice that the world is surprising [and] too promiscuous to fit easily or neatly into our stabilized linearities of change'. What our hope must do, is go to work. Our hope has to be active, not merely by existing, but by fuelling our diverse and unique approaches and dedication to change.

This book began with a reflection that our natural connection – cultural, historical, interpersonal and ecological – is inhibited or nurtured by the roots of our relationships with each other and the living world. That in order to usher in a safe, thriving and abundant future, we must move away from our obsession with individualism, guilt, perfectionism and saviourism towards an existence of interconnectedness, interdependence, diversity and kinship. The six sections of this book represent a complex web of stories, ideas and teachings that can ground us as we enter the unknown of a world facing climate and ecological breakdown. They are roots to hold on to and be held by. They have taught us that we must transform our rage into action and that there is power in resisting the logics of colonialism and capitalism that keep us from imagining more just futures. We have explored the need to place nature and the wisdom of the past at the centre of radical environmental innovations and how investing in theory is a means of generating new ideas to resist systems of oppression. Within these pages, we have recognised our collective grief as an opening to transform our connection to each other and the living world and that we must cultivate a universal ethic of care by looking to the planet's neglected nurturers. These words exist as reminders that so much of what we need to survive already exists. That we are more resourced, more powerful and wiser than we have been made to believe. That we come from a lineage of individuals deeply connected to and fiercely protective of the Earth, individuals who continue to live through us today. What I hope you have come to find is that there lies much more beyond the systems that dominate society and environmental action as we know it. That we have an opportunity to take what we need from the past and engage in an ongoing alchemical process of renewal and regeneration. Waiting not for the moment where everything is 'saved', 'fixed' or 'solved', but working towards repairing our hearts, our minds and our relationships with the living world and each other; by doing so, we open ourselves up to a beautiful mosaic of ways to be a part of positive change.

We have met and heard the stories of an inspiring collection of communities who are rooted in and demonstrate a natural connection with both people and planet in myriad ways. Many of these stories of natural connection hail from forests, rivers and valleys across the world but they hold wisdom that is essential for us all. They guide us in asking what we already are and can become native to. These stories are hope-bringers, but they are also change-makers. I share them in order to transform the way you engage in action and help you reimagine how it is you make change in the world. Your task now is to connect with these roots in your daily life and use them to build a natural connection with the world around you. Take these stories, the wisdom they impart, and write your own. Whether you are starting on a blank page, or on your fifth volume, take the stories in this book and use them to guide and nurture *your* natural connection. Reflect on the emotional wellsprings you can tap into to sustain your action. Which roots are you starving – maybe rage or grief or imagination? Which roots are you waiting for permission to put firmly in the ground – maybe innovation or theory? Which roots have you neglected? For many of us this will be radical care. Return to these stories and use them to help craft a sustainable practice of reflection and action. Sit in silence and meditate on the lessons of each section, grab a journal and map out a personal journey of integrating each root into your life, gather with your friends and your colleagues, and ask yourselves how you are cultivating a natural connection in your social and workspaces. However you carry these words with you in your daily life, remember that hope is not only a thing with feathers, as Emily Dickinson tells us, but hope also is a thing with

r

o

o

t

s.

Our hope must sing and soar and take to the wind, persisting despite the wildest storms, yet it must also be rooted, connected to the Earth and each other, stable and unwavering, in order to seed sustained change.

Acknowledgements

Writing this book, I have felt the presence of my eight-year-old and seventeen-year-old selves – the former full of curiosity and enthusiasm, the latter full of pain and fear – who would both be bowled over to know that, as they dreamt so many times, they are now a published author. I am immensely grateful for all who have come together, in many different ways, to make the dreams of those two girls a reality.

The beginnings of this book took root in the summer of 2021 when I met my wonderful agent Emma Leong, who not only encouraged me to embark on this journey but poured an immense amount of energy, care and guidance to bring these pages to life. Thank you, Emma, for your fierce and unwavering confidence in me as a writer, even when I have been unsure of it myself.

My deepest thanks to my editor Marianne Tatepo for believing in and moulding me as well as this book. Looking back at where this text started and then at the one we hold in our hands now, it is undeniable that your wisdom, vision and ambitions for this book have shaped and helped me grow immeasurably as a writer. Thank you to the rest of the team at Penguin Vintage and Square Peg – Emily Martin, Graeme Hall, Binita Roy, Mia Quibell-Smith, Matthew Broughton, Rowena Skelton-Wallace – who have read, edited, designed, marketed and made possible this book.

The heart of this book lies in the legacies, stories, victories, losses and lessons of the Indigenous and marginalised communities who have seeded positive change, shaped the environmental movement and transformed the world through their actions. It is through their enduring worldviews, practices and work that a natural connection emerges, for which I am eternally grateful. These stories are accompanied by an incredible constellation of interviews. I feel immensely privileged to have been able to connect and converse with so many individuals who inspire me and whose work I deeply respect. Thank you to all those who shared their time, thoughts and insights with me – Robert Macfarlane, Rebecca Solnit, Katherine May, Olafur Elliason, Merlin Sheldrake, Obatala Efunwale, Morningstar Khongthaw, Emmanuel Vaughan-Lee, Willow Defebaugh, Lucy Jones, Jessica J. Lee, Daze Aghaji, Báyò Akómoláfé, Sammy Oteng, Yvette Dickson-Tetteh, Adenike Oladosu, brontë velez, Mya-Rose Craig, Zena Holloway, Isaias Hernandez, Seth Hughes, Julia Watson, Miranda Lowe, Kalpana Arias – some of which made it into these pages and others that didn't make the final draft. I give thanks to all the skilled and inspiring writers who have come before me, whose pages I pored over and whose ideas have shaped the way I think about and move through the world.

So much of this book took shape in stolen moments, mornings, evenings, weekends, as I juggled writing with my PhD research, running ClimateInColour and ongoing commitments to my media and communications work. This book was born in the cracks between months-long field trips in the forests in Ghana, countless hours writing academic papers and my doctoral thesis and travelling for national and international engagements and appearances. This juggle would not have been possible without the patience, support and generosity of so many people. To my academic supervisors Alan Blackwell, Jennifer Gabrys, Adham Ashton-Butt and Emmanuel Acheampong, thank you so much for enabling me to undertake this work and for your mentorship, guidance and advocacy. To my manager Viv, thank you for your big heart, kindness and friendship and for steering the ship of my career. Thank you to my family, for their undying support and belief, for listening to me brainstorm, rant and vent, and for sharing their honest and helpful thoughts throughout the many stages of writing this book. Thanks especially to my mother for instilling in me, from an early age, my love for reading and writing and for being a model of determination, perseverance and grace that I aspire to every day. Thank you to my godmother, Aunty Augusta, for being such a steady and strong source of inspiration, motivation and wisdom. Thank you to Amelia, for your continual encouragement, for special days writing in your home by the river, and your ever-perceptive thoughts and comments.

And lastly, my love, Jasper. Thank you for your undying confidence and belief in me, for reading thousands of words countless times, for holding me and steadying me, for feeding and nourishing me, for championing me. For the natural connection we have fostered with each other over the last eight years. Thank you.

Select Bibliography

Aguon, Julian. 2022. *No Country for Eight-Spot Butterflies: A Lyric Essay.*

Albrecht, Glenn. 2019. *Earth Emotions: New Words for a New World.*

Armstrong, Karen. 2022. *Sacred Nature: How We Can Recover Our Bond with the Natural World.*

brown, adrienne maree. 2017. *Emergent Strategy: Shaping Change, Changing Worlds.*

brown, adrienne maree. 2019. *Pleasure Activism: The Politics of Feeling Good.*

Butler, Octavia E. 1993. *Parable of the Sower.*

Care Collective, The. 2020. *The Care Manifesto: The Politics of Interdependence.*

Davis, Angela, Y. 1981. *Women, Race and Class.*

Dungy, Camille. 2009. *Black Nature: Four Centuries of African American Nature Poetry.*

Fanon, Frantz. 1961. *Wretched of the Earth.*

Freire, Paulo. 1968. *Pedagogy of the Oppressed.*

Ghosh, Amitav. 2016. *The Great Derangement.*

Haraway, Donna J. 2016. *Staying with the Trouble: Making Kin in the Chthulucene.*

hooks, bell. 1999. *All about Love: New Visions.*

Johnson, G. T. and Lubin, A., eds. 2017. *Futures of Black Radicalism.*

Jones, Lucy. 2020. *Losing Eden: Our Fundamental Need for the Natural World and Its Ability to Heal Body and Soul.*

Kelley, Robin D. G. 2002. *Freedom Dreams: The Black Radical Imagination.*

Kimmerer, Robin Wall. 2013. *Braiding Sweetgrass: Indigenous Wisdom, Scientific Knowledge and the Teachings of Plants.*

Krawec, Patty. 2022. *Becoming Kin: An Indigenous Call to Unforgetting the Past and Reimagining Our Future.*

Okri, Ben. 2023. *Tiger Work: Stories, Essays and Poems about Climate Change.*

Page, Cara and Woodland, Erica. 2023. *Healing Justice Lineages: Dreaming at the Crossroads of Liberation, Collective Care, and Safety.*

Penniman, Leah. 2023. *Black Earth Wisdom: Soulful Conversations with Black Environmentalists.*

Piepzna-Samarasinha, Leah Lakshmi. 2018. *Care Work: Dreaming Disability Justice.*

Powers, Richard. 2018. *The Overstory.*

Riley, Cole Arthur. 2022. *This Here FLesh: Spirituality, Liberation, and the Stories that Make Us.*

Robinson, Kim Stanley. 2020. *The Ministry for the Future.*

Sheldrake, Merlin. 2020. *Entangled Life: How Fungi Make Our Worlds, Change Our Minds and Shape Our Futures.*

Shiva, Vandana. 1993. *Monocultures of the Mind: Perspectives on Biodiversity and Biotechnology.*

Solnit, Rebecca. 2004. *Hope in the Dark: Untold Histories, Wild Possibilities.*

Solnit, Rebecca. 2023. *Not Too Late: Changing the Climate Story from Despair to Possibility.*

Watson, J. 2019. *Lo-TEK: Design by Radical Indigenism.*

Wellcome Collection. 2022. *Rooted Beings.*

Wellcome Collection. 2022. *This Book is a Plant: How to Grow, Learn and Radically Engage with the Natural World.*

Wengrow, David and Graeber, David. 2021. *The Dawn of Everything: A New History of Humanity.*

Notes

PROLOGUE: TAKING ROOT

1. House, M. R. (n.d.). 'Devonian period'. Britannica.
2. Meyer-Berthaud, B., Scheckler, S. E., and Wendt, J. (1999). '*Archaeopteris* is the earliest known modern tree'. *Nature*, 398.
3. UC Museum of Paleontology (2011). 'The Devonian period'.
4. Normile, D. (2017). 'The world's first trees grew by splitting their guts'. Science Adviser.
5. National Geographic (n.d.). 'Devonian period'.
6. Ibid.
7. Algeo, T. J., and Scheckler, S. E. (2010). 'Land plant evolution and weathering rate changes in the Devonian'. *Journal of Earth Science, 21*(1); Smart, M. S., et al. (2022). 'Enhanced terrestrial nutrient release during the Devonian emergence and expansion of forests: Evidence from lacustrine phosphorus and geochemical records'. GSA Bulletin.
8. Thompson, J. (2023). 'Tree roots may have set off a mass extinction'. Scientific American.

INTRODUCTION: NATURAL CONNECTION

1. Krawec, P. (2022). *Becoming Kin: An Indigenous Call to Unforgetting the Past and Reimagining Our Future*. Nation Books.
2. Milan, S., and Treré, E. (2019). 'Big Data from the South(s): Beyond data universalism'. *Television & New Media*.
3. Nowell, C. E., and Magdoff, H. (2024, 6 September). 'Western colonialism'. Britannica.
4. Campbell-Stephens, R. (2020). 'Global majority: Decolonising the language and reframing the conversation about race'. Leeds Beckett University.

I RAGE IS RESISTANCE

1. Albrecht, G. (2019). *Earth Emotions: New Words for a New World*. Cornell University Press.

2. Stanley, S. K., Hogg, T. L., Leviston, Z., and Walker, I. (2021). 'From anger to action: Differential impacts of eco-anxiety, eco-depression, and eco-anger on climate action and well-being'. *The Journal of Climate Change and Health*, *1*(100003).
3. Miles, E. (2022). *Nature is a Human Right: Why We're Fighting for Green in a Grey World*. Dorling Kindersley.
4. Gregersen, T., Andersen, G., and Tvinnereim, E. (2023). 'The strength and content of climate anger'. *Global Environmental Change*, *82*, 102738.
5. Stanley, S. K., Hogg, T. L., Leviston, Z., and Walker, I. (2021). 'From anger to action: Differential impacts of eco-anxiety, eco-depression, and eco-anger on climate action and wellbeing'. *The Journal of Climate Change and Health*, *1*, 100003; Marczak, M., Winkowska, M., Chaton-Østlie, K., Morote Rios, R., and Klöckner, C. A. (2023). '"When I say I'm depressed, it's like anger." An exploration of the emotional landscape of climate change concern in Norway and its psychological, social and political implications'. *Emotion, Space and Society*, *46*.
6. Cavaliere, C. T., and Ingram, L. J. (2023). 'Climate change and anger: Misogyny and the dominant growth paradigm in tourism'. *Annals of Leisure Research*, *26*(3), 354–371.
7. *The Times* and *The Sunday Times* (2023). 'Just Stop Oil activist punched to the floor by angry motorist'. (2023, 19 July). [YouTube video].
8. Lorde, A. (1997). 'The uses of anger'. *Women's Studies Quarterly*, *25*(1/2), 278–285.
9. Matt, S. (2021, 2 July). 'Love and . . . rage?'. William Temple Foundation.
10. Brown, A. D. (2018). 'How the wonder of nature can inspire social justice activism'. *Yes Magazine*.
11. Ibid.

2 THE DAWN OF THE ENVIRONMENTAL JUSTICE MOVEMENT

1. Mayor of London (2012). 'Air quality in city of London: A guide for public health professionals'.

2. Ibid.; Mayor of London (2024). 'Mayor launches UK's first targeted air quality alert for healthcare professionals'.
3. Mayor of London (n.d.). 'Health and exposure to pollution'.
4. White, N. (2022, 17 November). 'Black and Asian children face higher asthma hospitalisations'. *Independent*.
5. Taylor, M. (2021, 11 March). 'London teenagers' road signs highlight effect of toxic air on people of colour'. *Guardian*.
6. Sheila McKechnie Foundation (2022). 'Choked up'.
7. Transport for London (n.d.). 'Red routes'.
8. Mayor of London (2023). 'New report reveals the transformational impact of the expanded Ultra Low Emission Zone so far'.
9. Estien, C. O., Wilkinson, C. E., Morello-Frosch, R., and Schell, C. J. (2024). 'Historical redlining is associated with disparities in environmental quality across California'. *Environmental Science & Technology Letters*, *11*, 54–59. Vermeer, D. (2021, 16 August). 'Redlining and environmental racism'. School for Environment and Sustainability, University of Michigan.
10. Fowler, R. (2023, 1 January). 'The Ugly History of Redlining: A Federal Policy "Full of Evil"'. Tennessee Bar Association.
11. Hall, B. (ed.) (1988). *Environmental Politics: Lessons from the Grassroots*. Available at https://aroadtowalk.com/wp-content/uploads/2022/07/wc-website-dumping-on-warren-county-1988.pdf.
12. Fears, D., and Dennis, B. (2021, 6 October). 'Environmental justice and race'. *Washington Post*.
13. Ibid.
14. Labalme, J. (1988). 'Dumping on Warren County'. In *Environmental Politics: Lessons from the Grassroots*, edited by Bob Hall. Available at https://aroadtowalk.com/wp-content/uploads/2022/07/wc-website-dumping-on-warren-county-1988.pdf; Fears, D., and Dennis, B. (2021, 6 October). 'Environmental justice and race'. *Washington Post*.
15. Exchange Project. (2006). 'Real people—real stories'. University of North Carolina at Chapel Hill.
16. Kutz, J. (2022, 28 September). 'As the EPA introduces environmental justice office, the "mother of the movement" remembers the Black women who led the battle'. The 19th.

17. Atwater, W. (2022, 21 September). 'Warren County commemorates 40 years of environmental justice struggle'. North Carolina Health News.
18. Fears, D., and Dennis, B. (2021, 6 October). 'Environmental justice and race'. *Washington Post*.
19. Lattimore, W. (2023). 'Honoring the mothers of environmental justice'. The Christian Century.
20. Kutz, J. (2022, 28 September). 'As the EPA introduces environmental justice office, the "mother of the movement" remembers the Black women who led the battle'. The 19th.
21. Atwater, W. (2022, 21 September). 'Warren County commemorates 40 years of environmental justice struggle'. North Carolina Health News.
22. Kutz, J. (2022, 28 September). 'As the EPA introduces environmental justice office, the "mother of the movement" remembers the Black women who led the battle'. The 19th.
23. Labalme, J. (2022, 30 September). 'From the Archives: Dumping on Warren County'. Facing South.
24. Hall, B. (1988). 'Environmental politics: Lessons from the grassroots'. Available at https://aroadtowalk.com/wp-content/uploads/2022/07/wc-website-dumping-on-warren-county-1988.pdf.
25. Hall, B. (1988). Environmental politics: Lessons from the grassroots. *(No Title)*.
26. Lazarus, R. J. (2000). 'Environmental racism! That's What It Is'. *University of Illinois Law Review*, 255–274.
27. Office of Legacy Management (n.d.). 'History of environmental justice at the Department of Energy'.
28. Marsh, B. (2022, 28 September). '"A much-needed step": The EPA creates a new environmental justice office'. Grist.
29. Labalme, J. (2022, 30 September). 'From the Archives: Dumping on Warren County'. Facing South.
30. Sanusi, T. (2021, 8 November). 'Lake Chad is drying up. Meet the Nigerian activist fighting to save the lake and its people'. Global Citizen.

3 THE OGONI 9: NIGERIA'S FIGHT AGAINST FOSSIL FUELS

1. Kenneth, M. (2021, 9 November). 'A wealth of sorrow: Why Nigeria's abundant oil reserves are really a curse'. *Guardian*.
2. Kaamil A. (2021, 29 Jan) '"Finally some justice": Court Rules Shell Nigeria must pay for oil damage'. *Guardian*.
3. BBC News (11 August 2021). 'Shell pays $111m over 1970s oil spill in Nigeria'.
4. Pegg, S., and Zabbey, N. (2013). 'Oil and water: The Bodo spills and the destruction of traditional livelihood structures in the Niger Delta'. *Community Development Journal, 48*(3), 391–405.
5. Osaghae, E. E. (1995). 'The Ogoni uprising: Oil politics, minority agitation and the future of the Nigerian state'. *African Affairs, 94*(376), 325–344.
6. Amnesty International (2017, 29 June). 'Nigeria: Shell complicit in the arbitrary executions of Ogoni Nine as writ served in Dutch court'.
7. Ibid.
8. Ibid.
9. Ekpali, S. (2023, 12 December). 'The Niger Delta community devastated by yet another Shell oil spill'. Open Democracy.
10. Ogoninews.com. (2022). '"I'll tell you this, I may be dead but my ideas will not die." – Ken Saro-Wiwa, 1995'.
11. Friends of the Earth (2021, 29 January). 'Court orders Shell to pay Nigerian farmers over oil spills'.
12. Reuters (2022, 23 December). 'Shell to pay 15mln euros in settlement over Nigerian oil spills'.
13. Friends of the Earth (2021, 29 January). 'Court orders Shell to pay Nigerian farmers over oil spills'.

4 FOLLOWING RIVERS OF RESISTANCE: THE QUILOMBOLA FIGHT IN BRAZIL

1. Miki, Y. (2012). 'Fleeing into slavery: The insurgent geographies of Brazilian quilombolas (Maroons), 1880–1881'. *The Americas, 68*(4), 495–528.

2. Olympics.com. (2016, 06 December). 'How do we know that Rio 2016 was a success?'
3. Gross, D., and Watts, J. (2016, 21 July). 'Olympics media village built on "sacred" mass grave of African slaves'. *Guardian*.
4. Brum, E. (2020, 5 March). 'Fierce Life: Maria do Socorro Silva'. *Atmos*.
5. Nicas, J., and Milhorance, F. (2024, 24 March). 'Police say they've cracked Rio de Janeiro's most notorious murder mystery'. *New York Times*.
6. Dávila, S. A. (2023). 'How many more Brazilian environmental defenders have to perish before we act? President Lula's challenge to protect environmental Quilombola defenders'. *William & Mary Environmental Law and Policy Review*, 47(3), 657.
7. Ibid.
8. Watts, J. (2018, 21 June). '"They should be put in prison": Battling Brazil's huge alumina plant'. *Guardian*.

5 THE ORIGINAL TREEHUGGERS: INDIA'S ANTI-DEFORESTATION MOVEMENT

1. Biswas, S. (2021, 21 May). 'Sunderlal Bahuguna: The man who taught India to hug trees'. BBC News.
2. Baker, D. (1984). '"A serious time": Forest satyagraha in Madhya Pradesh, 1930'. *The Indian Economic & Social History Review*, 21(1), 71–90.
3. Halder, T. (2021, 7 April). 'Coming to terms with Gandhi's complicated legacy'. Al Jazeera.
4. Shiva, V., and Bandyopadhyay, J. (1986). 'The evolution, structure, and impact of the Chipko movement'. *Mountain Research and Development*, 6(2), 133–142.
5. Biswas, S. (2021, 21 May). 'Sunderlal Bahuguna: The man who taught India to hug trees'. BBC News.
6. War Resisters' International (2001, 1 January). 'The Chipko Movement'.
7. Ibid.
8. Biswas, S. (2021, 21 May). 'Sunderlal Bahuguna: The man who taught India to hug trees'. BBC News.

9. Bedi, S. (2022, 18 April). 'The Chipko movement: Treehuggers of India'. The Nonviolence Project.
10. Mountain Shepherds (n.d.). 'Chipko heritage'.
11. Ibid.
12. Bandyopadhyay, J. (1999). 'Chipko movement: Of floated myths and flouted realities'. *Economic and Political Weekly, 34*(15), 880–882.
13. Nabourema, F. (2021, 3 March). 'In India, women propel world's largest protest movement'. United States Institute of Peace.
14. Bhowmick, N. (2021, 4 March). '"I cannot be intimidated. I cannot be bought." The women leading India's farmers' protests'. *Time*.
15. Ibid.
16. The Commons Social Change Library (n.d.). 'The four roles of social activism by Bill Moyer'.
17. Building Movement Project (n.d.). 'Social change ecosystem map'.
18. Slow Factory (n.d.). 'Callings and roles for collective liberation'.
19. Gregory, T. (n.d.). 'Climate action roles poster download'.

6 JOY IS THE SISTER OF RAGE

1. Walsh, A. (2022, 29 May). 'Eco-fascism: The greenwashing of the far right'. DW.
2. Hickel, J. (2020). 'Quantifying national responsibility for climate breakdown: An equality-based attribution approach for carbon dioxide emissions in excess of the planetary boundary'. *The Lancet Planetary Health, 4*(9), e399–e404.
3. Lanham, J. D. (2022, 31 August). 'Joy is the justice we give ourselves'. *Emergence Magazine*.

8 RECLAIMING OUR COLLECTIVE IMAGINATION

1. All About Birds, Cornell Lab for Ornithology (n.d.). 'Clark's nutcracker'.
2. Jones, C. (2013, 3 December). '2 species found that owe lives to Giant Sequoias'. SFGate.
3. Rainforest Alliance (2023, 13 September). 'Kapok tree'.

4. Marshmallow Laser Feast (n.d.). 'Works of nature'.
5. Ibid.
6. Chevalier, C. (2016, 28 December). 'How this group of tree huggers is using VR to inspire conservation'. VRScout.
7. Ghosh, A. (2018). *The Great Derangement: Climate Change and the Unthinkable*. Penguin UK.
8. Naidoo, K. (2022, 4 November). 'An open letter to the philanthropic community: Harnessing the power of arts and culture for humanity's survival'.

9 DECOLONISING IMAGINATION

1. African American Policy Forum (2020, 5 August). 'Under the Blacklight: Storytelling while Black and female: Conjuring beautiful experiments'. [YouTube video].
2. Fecht, S. (2022, 22 September). 'Mining, land grabs, and more: When decarbonization conflicts with human rights'. Columbia Climate School, State of the Planet.
3. Krawec, P. (2022). *Becoming Kin: An Indigenous Call to Unforgetting the Past and Reimagining Our Future*. Nation Books.
4. de Saint-Laurent, C. (2018). 'Thinking through time: From collective memories to collective futures'. In *Imagining Collective Futures: Perspectives from Social, Cultural and Political Psychology*, edited by Constance de Saint-Laurent, Sandra Obradović and Kevin R. Carriere (pp. 59–81).
5. Veltman, C. (2023, 19 September). 'A Northern California tribe works to protect traditions in a warming world'. NPR.
6. National Park Service (2024, 18 March). 'Indigenous fire practices shape our land.'
7. Nazaryan, A. (2016, 17 August). 'Newsweek: "California Slaughter: The State-Sanctioned Genocide of Native Americans"'.
8. Amaral-Rogers, V. (2017, 27 January). 'Why nightingales matter'. RSPB.
9. Wright, G. (2022, 29 November). *Cailleach*. Mythopedia; Blackie, S. (2019, 25 October). 'How Irish myth and folklore can inspire women to fight for ecological change'. *Irish Times*.

10. Haiven, M. (2022, 17 June). 'Dreaming together: Artists mobilizing collective dreaming methods for the radical imagination (Capacious)'.

10 DOES A GLACIER MOURN ITS DEATH?

1. Reid, G. (2022, 29 April). 'Welcome to the Planthroposcene: A conversation with Natasha Myers'. *Wonderground*, Issue 1, Conversations, Culture.
2. Allison, E. A. (2015). 'The spiritual significance of glaciers in an age of climate change'. *Wiley Interdisciplinary Reviews: Climate Change*, 6(5), 493–508.
3. Bressan, D. (2021, 21 July). 'Melting glaciers reveal previously unknown viruses'. Forbes.
4. Zavalkoff, A. (2004). 'Dis/located in nature? A feminist critique of David Abram'. *Ethics and the Environment*, 9(1), 121–139.

11 PARABLES FOR THE FUTURE

1. Hudson, A. J. (2021). 'The end of the world, for whom? or, whose world? Whose ending? An Afrofuturist and Afropessimist counter perspective on climate apocalypse'. *American Studies*, 60(3/4).
2. Reuters (2024, 10 July). 'Gaza death toll: How many Palestinians has Israel's campaign killed?'
3. Operation Broken Silence (2024, 2 September). 'Sudan crisis 2024: What you need to know'.
4. Aguon, J. (2022). *No Country for Eight-Spot Butterflies: A Lyric Essay*. Astra Publishing House.
5. Mbewe, M. (2020, 3 December). 'What Afrofuturism teaches us about environmental feminism'. Sister Namibia.
6. Halstead J. (2022, 5 October). 'The most dangerous story ever told: Ecological collapse, progress, and human destiny'. Medium.
7. Mapondera, M. (2020). 'If another world is possible, who is doing the imagining? Building an ecofeminist development alternative in a time of deep systemic crisis'. African Women's Development Fund.
8. Nnedi Okorafor (2019, 15 October). 'Africanfuturism defined'.

9. Canavan, G. (2014, 9 June). '"There's nothing new / Under the sun, / But there are new suns": Recovering Octavia E. Butler's Lost Parables'. LA Review of Books.
10. Twahirwa, R. P. (2022). 'On the destiny of our species: Reading Octavia E. Butler's Parables series'. LSE Review of Books.

13 (HOW) WILL TECHNOLOGY SAVE THE PLANET?

1. UNFCCC (2022, 21 October). 'Innovation Hub at COP27 to promote transformative climate solutions'.
2. Association for British Mining Energy and Consultancy (2022, 8 June). 'Egypt commits $40bn to green hydrogen economy to attract foreign investment'.
3. O'Farrel, S. (2022, 10 October). 'Egypt emerges unlikely green energy powerhouse'. FDI Intelligence.
4. Levin, K., and Steer, A. (2021, September). 'Fighting climate change with innovation'. International Monetary Fund.
5. CTVC (2022, 15 July). 'Climate funding abides in $19bn 2022 midyear update'.
6. Almond, L. (2022, September). 'What you can measure, you can manage – how Nature Tech can help us solve the nature and climate crises'. Nature4Climate.
7. ClimateTrade (2022, 4 August). 'Voluntary Carbon Market value tops $2bn'.
8. Almond, L. (2022, September). 'What you can measure, you can manage – how Nature Tech can help us solve the nature and climate crises'. Nature4Climate.
9. University of Bath (2021, 25 November). '£17M iCAST project launches at Swindon Carriage Works'.
10. Adler, T. (2021, 15 May). 'British investors ready to invest £1.4bn in university spinouts'. Growth Business.
11. University of Sussex (2022). 'Thirty years of climate research funding has overlooked the potential of experimental transformative technologies'.
12. Hao, K. (2020, 4 December). 'We read the paper that forced Timnit Gebru out of Google. Here's what it says'. MIT Technology Review.

13. Ferreira Santos, S. (2024, 11 August). 'Thousands protest against lithium mining in Serbia'. BBC News.
14. Blok, V. (2021). 'Philosophical reflections on the concept of innovation'. In *Handbook on Alternative Theories of Innovation* (pp. 354–367). Edward Elgar Publishing.
15. National Geographic Education (n.d.). 'Industrial revolution and technology' (7th grade).
16. Ibid.
17. Heblich, S., Redding, S. J., and Voth, H. J. (2022). 'Slavery and the British industrial revolution'.
18. Arnold, D. (2005). 'Europe, technology, and colonialism in the 20th century'. *History and Technology*, *21*(1), 85–106; Bang, M., Marin, A., Faber, L., and Suzukovich III, E. S. (2013). 'Repatriating indigenous technologies in an urban Indian community'. *Urban Education*, *48*(5), 705–733.
19. Osuala, U. S. (2012). 'Colonialism and the disintegration of indigenous technology in Igboland: A case study of blacksmithing in Nkwerre'. *Historical Research Letter*, *3*, 11–19.
20. Jarwar, M. A., Dumontet, S., and Pasquale, V. (2024). 'The natural world in Western thought'. *Challenges*, *15*(1), 17.
21. University of Winnipeg (n.d.). 'Justice Murray Sinclair'.
22. Archie K., and Bolduc, J. (2018). 'An invitation to explore indigenous innovation'. *Stanford Social Innovation Review*.

14 WAKANDAN COSMOLOGY: A BLUEPRINT FOR ROOTED INNOVATION

1. Akhalbey, F. (2019, Match 24). 'Kwame Nkrumah's iconic 1963 speech on African unity'. Face2Face Africa.
2. Nkrumah, K. (1967). 'African socialism revisited'.
3. Maryville University (2020, 9 January). 'Women of color in STEM'.
4. Rankin, Y. A., and Thomas, J. O. (2020, February). 'The intersectional experiences of Black women in computing'. In *Proceedings of the 51st ACM technical symposium on computer science education* (pp. 199–205).
5. Simonite, T. (2020, 7 December). 'What really happened with Google's Timnit Gebru'. Wired.

6. Ojewale, O. (n.d.). 'Child miners: The dark side of the DRC's coltan wealth'. Institute for Security Studies.
7. Bracking, S. (2009). 'Hiding conflict over industry returns: A stakeholder analysis of the Extractive Industries Transparency Initiative'. Global Development Institute Working Paper Series 9109, GDI, The University of Manchester.
8. Kennedy, R. (2018, 28 June). 'Society, technology, and cosmology in Black Panther'. In *Anthropology of Black Panther*.
9. Flatow, N. (2018, 5 November). 'How Hannah Beachler built Black Panther's Wakanda'. Bloomberg.
10. Hadadi, R. (2022, 22 November). 'Hannah Beachler on the "Wakanda Forever" set design and Easter eggs. Vulture.
11. Vemula, V. (n.d.). 'An architectural review of Black Panther'. Rethinking The Future.
12. Sam, S., et al. (2022). 'Co-designing a platform for documenting African Indigenous knowledge: Participatory citizen science and data science approaches'. In *Global Transformations in Media and Communication Research 2022* (pp. 1–10). Palgrave Macmillan.
13. Kalele, D. N., Ogara, W. O., Oludhe, C., and Onono, J. O. (2021). 'Climate change impacts and relevance of smallholder farmers' response in arid and semi-arid lands in Kenya'. *Scientific African*, *12*, e00814; Parracciani, C., Buitenwerf, R., and Svenning, J. C. (2023). 'Impacts of climate change on vegetation in Kenya: Future projections and implications for protected areas'. *Land*, *12*(11), 2052.
14. World Weather Attribution (2023, 27 April). 'Human-induced climate change increased drought severity in the southern Horn of Africa'.
15. Farm to Market Alliance (2022). 'Making markets work better for smallholder farmers: FtMA Kenya country brief'.
16. Masinde, M., and Bagula, A. (2012). 'ITIKI: Bridge between African indigenous knowledge and modern science of drought prediction'. *Knowledge Management for Development Journal*, *7*(3).
17. Ibid.
18. Central University of Technology, Free State (2021, 18 February). 'Prof. Muthoni Masinde presents to Scifest Africa on application of Design Thinking in the creation of ITIKI tool'.

19. Arnold, D. (2005). 'Europe, technology, and colonialism in the 20th century'. *History and Technology*, *21*(1), 85–106.

15 ROOTS OF INNOVATION

1. Meghalaya Tourism (n.d.). 'About Meghalaya'.
2. India Trail (n.d.). 'Meghalaya'.
3. Lyngkhoi, W. (2022). 'Living root bridges: A case study of Amdohkha, Amtren, and Amkhlew in Nongtalang, Meghalaya'. *International Journal of Novel Research and Development*, *7*(5), 290–295.
4. Watson, J. (2019). *Lo-TEK: Design by Radical Indigenism*. Taschen.
5. Shankar, S. (2015, September). 'Living root bridges: State of knowledge, fundamental research and future application'. In *Proc. of 2015 IABSE Conf.—Structural Engineering: Providing Solutions to Global Challenges*, *105*, pp. 1–8.
6. Middleton, W., Habibi, A., Shankar, S., and Ludwig, F. (2020). 'Characterizing regenerative aspects of living root bridges'. *Sustainability*, *12*(8), 3267.
7. Shankar, S. (2016, 5 August). 'Route by root'. *Down To Earth Magazine*.
8. Imphal Free Press Bureau (2023, 7 March). 'Stop false projection of large dams as climate change-friendly: Northeast CSOs'.
9. Hub Network (2023, 13 July). 'Illegal mining behind presence of illegal coke factories, continued drone survey needed to check it, says justice Katakey'.
10. Substack (2022, 18 July). *Diengiei: The Terrible Tree of Despair*.
11. Chowdhury, P. D. (2020). '"We are a story-telling people"': An exploration of the origin tales in the oral narratives of the Nagas and the Khasis of India's North East. *Towards Excellence*, *12*(2).
12. Royal Institute of British Architects (2023, 2 May). 'Built for the environment: Report'.

16 THE FUTURE LIES IN THE CLAY: IRAN'S ENERGY-FREE COOLING SOLUTIONS

1. Gowen, A. et al. (2023, 22 September). 'The inequality of heat'. *The Washington Post*.
2. Kumar, A. (2023, 2 June). 'Delhi's temperature did not cross 50 degrees; it almost touched it, says IMD'. *Business Standard*.
3. AuthIndia (2022, 23 September). 'CoolAnt zero-energy air conditioner: A "terracotta" air conditioner designed by Ant Studio'.
4. Ibid.
5. United Nations Environment Programme (2019, 4 February). 'Indian architect turns bees and terracotta to design innovative cooling system.'
6. Shokoohi, K. (2021, 10 August). 'The ancient Persian way to keep cool'. BBC.
7. Ibid.
8. Bouscaren, D. (2022, 14 September). 'Long before electricity, wind catchers of Persia kept residents cool. Climate-conscious architects are taking notes'. The World.
9. Material Cultures (n.d.). 'Waste Age 2021'.
10. United Nations Environment Programme (2023, 12 September). 'Building materials and the climate: Constructing a new future'.
11. House of Commons Environmental Audit Committee (2022). 'Building to net zero: Costing carbon in construction'. First Report of Session 2022–23.
12. Clay Brick Association (n.d.). 'Factsheet #07: The lasting legacy of clay brick'.
13. Natural Homes (n.d.). 'The natural vernacular architecture of Africa'.
14. Siyabona Africa, Kruger Park (n.d.). 'San: History of the San people'.
15. Africa Freak (n.d.). 'The San people of Africa - Guide to the Kalahari Bushmen Tribes'.
16. Siyabona Africa, Kruger Park (n.d.). 'San: History of the San people'.
17. Ibid.
18. Survival International (2006, 13 December). 'Bushmen win landmark legal case'.

18 THEORY AS LIBERATION

1. hooks, b. (1991). 'Theory as liberatory practice'. *Yale Journal of Law & Feminism*, 4(1).
2. Critical Review (2018, 22 March). 'Angela Davis, "We Need Intellectuals"'. [YouTube video].

19 IT TAKES A LAWYER, AN ACTIVIST AND A STORYTELLER (AND YOU) TO CHANGE THE WORLD

1. Gevisser, M., and Redford, K. (2023). *The Revolution Will Not Be Litigated: People Power and Legal Power in the 21st Century*. OR Books.
2. Ibid.
3. Bullard, R. D. (2018). *Dumping in Dixie: Race, Class, and Environmental Quality*. Routledge.
4. Reuters (2023, 5 May). 'U.S. settles landmark Alabama environmental justice case'.
5. NPR (2023, 11 October). 'Landmark environmental justice case aims to fix an Alabama county's sewage problems'.
6. Office of Public Affairs, US Department of Justice (2023). 'Departments of Justice and Health and Human Services announce interim resolution agreement in environmental justice investigation of Alabama Department of Public Health'.
7. US Department of Justice (2023). 'Comprehensive environmental justice enforcement strategy: Annual report'.

20 THEORY AS COLLECTIVE WISDOM

1. National Library of New Zealand (n.d.). 'Change-maker – the Whanganui River'.
2. Te Ara (n.d.). *Story: Whakapapa – genealogy*.
3. Kramm, M. (2020). 'When a river becomes a person'. *Journal of Human Development and Capabilities*, 21(4), 307–319.

4. Ngā Tāngata Tiaki o Whanganui (n.d.). 'Our story: Te Awa Tupua'.
5. Whanganui District Council (n.d.). 'Te Awa Tupua - Whanganui River settlement'.
6. West Coast Environmental Law (n.d.). 'I am the River and the River is me: Legal personhood and emerging rights of nature'.
7. Stone, C. D. (2017). 'Should trees have standing? Toward legal rights for natural objects'. In *Environmental Rights*, edited by Steve Vanderheiden (pp. 283–334). Routledge.
8. Johnson, M. (2017, 14 June). 'The river is not a person: Indigeneity and the sacred in Aotearoa New Zealand'. The Immanent Frame – Social Science Research Council.
9. Ngā Tāngata Tiaki o Whanganui (n.d.). 'Our story: Te Awa Tupua'.
10. Environment Institute of Australia and New Zealand Inc. (n.d.). 'Gerrard Albert & Dr. Nic Peet'.
11. Takacs, D. (2021). 'We are the river'. *University of Illinois Law Review*, 2021(2), 545.
12. Rasch, E. D., van der Hout, F., and Köhne, M. (2022). 'Engaged anthropology and scholar activism: Double contentions'. *Contention*, 10(1), 1–12.

21 STRADDLING WORLDS FOR RESISTANCE AND CHANGE

1. *Financial Times* (2023, 17 July). 'Cambridge University should halt funding from fossil fuel groups, report finds'.
2. hooks, b. (1991). 'Theory as liberatory practice'. *Yale Journal of Law & Feminism*, 4(1).
3. Lipsitz, G. (2017). 'What is the Black Radical Tradition?' In *Futures of Black Radicalism*, edited by G. T. Johnson and A. Lubin (pp. 17–36). Verso Books.
4. hooks, b. (2006). *Outlaw culture: Resisting Representations*. Routledge.
5. Myers, J. (2017). 'Cedric Robinson and the Black Radical Tradition'.
6. Johnson, G. T., and Lubin, A. (eds). (2017). *Futures of Black Radicalism*. Verso Books.

7. hooks, b. (2014). *Teaching to Transgress*. Routledge.
8. The Goldman Environmental Prize (n.d.). 'Sharon Lavigne'.
9. Forensic Architecture (2021). 'Environmental racism in Death Alley, Louisiana: Phase 1 report'.
10. Lartey, J., and Loughland, O. (2019, 6 May). 'Cancertown: "Almost every household has someone that has died from cancer"'. *Guardian*.
11. Forensic Architecture (2021). 'Environmental racism in Death Alley, Louisiana: Phase 1 report'.
12. Ibid.
13. The Goldman Environmental Prize (n.d.). 'Sharon Lavigne'.

22 SYSTEMS CHANGE, NOT CLIMATE CHANGE

1. Dundee Glacier (n.d.). 'The Tay: UK's largest river'; McKinsey & Company. (2014, 1 February). 'Navigating the circular economy: A conversation with Dame Ellen MacArthur'.
2. McKinsey & Company (2014, 1 February). 'Navigating the circular economy: A conversation with Dame Ellen MacArthur'.
3. Bottle, R., and Gu, J. (2022, 23 February). 'The Real Environmental Impact of the Fashion Industry'. Bloomberg; World Bank Group. (2019, 23 September). 'How much do our wardrobes cost to the environment?'
4. Ruiz, A. (2024, 18 March). '17 Most Worrying Textile Waste Statistics & Facts'. The Roundup; Robinson, F. (2024). 'No One Knows How Many Clothes Are Made. Why Won't Brands Tell Us?'. Good on you.
5. Rudgard, O. (2022, 2 September). 'Oxfam's unwanted clothes are washing up as rubbish in Africa'. *Telegraph*.
6. Benjamin, B. (2022). 'The second-hand market at the heart of Ghana's fashion revolution'. Dazed Digital.
7. Britten, F. (2022, 8 June). 'Fast fashion giant Shein pledges $15m for textile waste workers in Ghana'. *Guardian*.
8. Jin, L., et al. (2018). 'Modeling future flows of the Volta River system: Impacts of climate change and socio-economic changes'. *Science of the Total Environment*, 637, 1069–1080.

24 THE EARTH IS A CHURCH: ETHIOPIA'S ARCHITECTURE-INSPIRED CONSERVATION

1. Crummy, D., and Mehretu A. (2024, 11 August). 'Religion of Ethiopia'. Britannica.
2. Mosissa, D., and Abraha, B. (2018). 'A review of conservation of biodiversity in sacred natural sites in Ethiopia: The role of Ethiopian Orthodox Tewahedo Church'. *Journal of Plant Science Research*, 5(1), 176.
3. Eshete, A. W. (2007). 'Ethiopian church forests: Opportunities and challenges for restoration'. Wageningen University and Research.
4. Ibid.
5. *Emergence Magazine* (2020, 3 January). 'The church forests of Ethiopia'.
6. Ibid.
7. Eshete, A. W. (2007). 'Ethiopian church forests: Opportunities and challenges for restoration'. Wageningen University and Research.
8. Ethiopian Orthodox Church (n.d.). 'Worship in the Ethiopian Orthodox Church'.

25 NATURE IS A HUMAN RIGHT

1. Wellcome (n.d.). 'Wellcome photography prize 2021: Managing mental health'.
2. Lowry, C. A., et al. (2007). 'Identification of an immune-responsive mesolimbocortical serotonergic system: Potential role in regulation of emotional behavior'. *Neuroscience, 146*(2), 756–772.
3. University of Essex (2022, 3 November). 'Gardening eased lockdown loneliness as pandemic hit'.
4. Alvarsson, J. J., Wiens, S., and Nilsson, M. E. (2010). 'Stress recovery during exposure to nature sound and environmental noise'. *International Journal of Environmental Research and Public Health, 7*(3), 1036–1046.
5. Forest Therapy Hub (n.d.) 'What is forest therapy'.

6. Rosa, C. D., Larson, L. R., Collado, S., and Profice, C. C. (2021). 'Forest therapy can prevent and treat depression: Evidence from meta-analyses'. *Urban Forestry & Urban Greening, 57*, 126943.
7. Miyazaki, Y. (2021). *Walking in the Woods: Go back to Nature with the Japanese Way of Shinrin-yoku*. Aster.
8. Ibid.
9. Rots, A. P. (2015). 'Sacred forests, sacred nation: The Shinto environmentalist paradigm and the rediscovery of "chinju no mori"'. *Japanese Journal of Religious Studies, 42*(2), 205–233.
10. Ibid.
11. Tsui, T. (2023). *It's not just you: How to navigate eco-anxiety and the climate crisis*. Simon & Schuster.
12. Horton, H. (2023, 2 March). 'Nearly half of English neighbourhoods have less than 10% tree cover'. *Guardian*.
13. ABC News (2023, 16 July). 'Extreme heat sweeps the world from Europe to the US and Japan'; Simón, Y. (2024, 26 April). 'The 10 hottest states in the U.S., based on 2023 data'. howstuffworks; Weather&Radar (2023, 18 July). 'China shatters all-time high: 52.2°C temperature record'.
14. New Scientist (2021, 23 November). 'Trees cool the land surface temperature of cities by up to 12°C'.
15. Groundwork (2021). 'Out of bounds'.
16. Spotswood, E. N., et al. (2021). 'Nature inequity and higher COVID-19 case rates in less-green neighbourhoods in the United States'. *Nature Sustainability, 4*(12), 1092–1098.
17. Groundwork (2021, 11 May). 'News: Report finds severe inequalities in access to parks and greenspaces in communities across the UK'.
18. NPR (2023, 29 June). 'Central Park birder Christian Cooper on being a Black man in the natural world'.

26 HEALING THROUGH THE CRACKS

1. Marczak, M., et al. (2023). '"When I say I'm depressed, it's like anger." An exploration of the emotional landscape of climate change concern in Norway and its psychological,

social and political implications'. *Emotion, Space and Society*, 46.
2. The Bureau of Linguistical Reality (n.d.). 'Solastalgia'; Albrecht, G. et al. (2007). 'Solastalgia: The distress caused by environmental change'. *Australasian Psychiatry*, 15(1), S95–S98.
3. The Kipling Society (n.d.). 'The white man's burden'.
4. Emory Scholar Blogs (n.d.). 'The philosophy of colonialism: Civilization, Christianity, and commerce'.
5. University College London News (2019, 27 February). '"Great Dying" in the Americas disturbed Earth's climate'.
6. Curtin, P. D. (1964). *The Image of Africa: British Ideas and Action, 1780–1850*. University of Wisconsin Press.
7. Mahony, M., and Endfield, G. (2018). 'Climate and colonialism'. *Wiley Interdisciplinary Reviews: Climate Change*, 9(2), e510.
8. Livingstone, D. N. (2005). 'Scientific enquiry and the missionary enterprise'. In *Participating in the Knowledge Society: Researchers Beyond the University Walls*, edited by R. Finnegan (pp. 50–64). Palgrave.
9. Descartes, R., and Johnston, I. C. (2003). *Discourse on Method for Reasoning Well and for Seeking Truth in the Sciences*. Prideaux Street Publications.
10. Center for Sustainable Systems, University of Michigan (n.d.). 'U.S. renewable energy factsheet'; The Big Issue (n.d.). 'Renewable energy: How does the UK compare to other countries?'; Kleinman Center for Energy Policy (n.d.). 'Kenya's clean energy transition gets a boost from solar power'.
11. Encyclopedia Virginia (n.d.). 'Slave ships and the Middle Passage'; Royal Museums Greenwich (n.d.). 'Dying on their own terms: Suicides aboard slave ships'.

27 GRIEF IS THE WAY TO TRANSFORMATION

1. For the wild (2023, 19 July). 'The Edges in the Middle, V: Báyò Akómoláfé, Naomi Klein, and Yuria Celidwen'.
2. CNN (2017, 29 December). 'Celebrating death in style: Ghana's fantasy coffins'.
3. Trees for Life. (n.d.). 'Yew: Mythology and folklore of yew'.

4. Starling, H. (2023). *The Bleeding Tree: A Pathway through Grief Guided by Forests, Folk Tales and the Ritual Year*. Random House.
5. velez, b. (n.d.). 'Lead to life: The alchemy of atonement'. The Gia Reader.
6. Ibid.
7. Steinauer-Scudder, C. (2018, 12 September). 'Applied alchemy'. *Emergence Magazine*.
8. velez, b. (n.d.). 'Lead to life: The alchemy of atonement'. The Gia Reader.

29 WE ARE ALL WE HAVE

1. Piepzna-Samarasinha, L. L. (2018). *Care Work: Dreaming Disability Justice*. Arsenal Pulp Press.
2. Gilligan, C. (1982). *In a Different Voice: Psychological Theory and Women's Development*. Harvard University Press; Held, V. (2006). *The Ethics of Care: Personal, Political, and Global*. Oxford University Press; Noddings, N. (2013). *Caring: A Relational Approach to Ethics and Moral Education* (2nd ed.). University of California Press.
3. Francis-Devine, B. (2022). 'Poverty in the UK: Statistics'.
4. Trussell Trust (n.d.). 'Food bank statistics 2022–2023'.
5. PA Media (2023, 11 January). 'Alarm raised at decline in women's maternity experiences in England'. *Guardian*.
6. Engage Britain (n.d.). 'Millions suffer poor care'; Kirkup, J. (2023, 15 June). 'Failure of the care system should shame us all'. *The Times*.
7. Benefits and Work (n.d.). 'Tougher sanctions target claimants: Cash, medication, and access to justice'.
8. Advice Now (n.d.). 'Avoid or challenge a universal credit sanction'.
9. The Care Collective (2020). *The Care Manifesto: The Politics of Interdependence* (p. 17).
10. Lisa Chamberlain, 'From self-care to collective Care'. *SUR* 30 (2020).
11. Lorde, A. (2017). *A Burst of Light: And Other Essays*. Courier Dover Publications.

12. The Care Collective (2020). *The Care Manifesto: The Politics of Interdependence* (p. 22).
13. Council on Foreign Relations (2024, 20 June). 'Conflict in the Democratic Republic of Congo'.
14. Zuckerman, J. (2023, 30 March). 'For your phone and EV, a cobalt supply chain to a hell on Earth'. Yale E360.
15. National Geographic Education (n.d.). 'Anthropocene'.
16. The Care Collective (2020). *The Care Manifesto: The Politics of Interdependence* (p. 5).

30 BEYOND THE BURDEN OF CLIMATE CARE

1. Moriggi, A., Soini, K., Franklin, A., and Roep, D. (2020). 'A care-based approach to transformative change: Ethically-informed practices, relational response-ability & emotional awareness'. *Ethics, Policy & Environment, 23*(3), 281–298.
2. Kaufman, M. (n.d.). 'The carbon footprint sham'. Mashable; DeSmog (n.d.). 'Ogilvy'.
3. Seager, A., and Bowers, S. (2006, 21 April). 'BP attacked over CO2 emissions'. *Guardian*.
4. LSE Blogs (2022, 24 January). 'Embodying collective care through decolonial feminist praxis'.
5. Kashwan, P. (2017). 'Inequality, democracy, and the environment: A cross-national analysis'. *Ecological Economics, 131*, 139–151.
6. MacKillican, A. C. (2022). 'The red deal by the red nation'. *Journal of Multidisciplinary Research at Trent, 3*(1), 155–157.
7. Statista (n.d.). 'Indigenous communities protect 80% of all biodiversity'.
8. FAO and FILAC (2021). 'Forest governance by indigenous and tribal peoples. An opportunity for climate action in Latin America and the Caribbean'.
9. Piepzna-Samarasinha, L. L. (2018). *Care Work: Dreaming Disability Justice*. Arsenal Pulp Press.
10. Ibid. (p. 17).

31 CLIMATE WORK IS CARE WORK

1. Lakhani, N. (2023, 5 December). 'Record number of fossil fuel lobbyists get access to COP28 climate talks'. *Guardian*.
2. Piepzna-Samarasinha, L. L. (2018). *Care Work: Dreaming Disability Justice*. Arsenal Pulp Press (p. 19).
3. Kauffman, J. (2017, 6 November). 'Teresa Santos: Died in Napa fire providing care for an older woman'. SF Gate.
4. The Leap (2020). 'Care work is essential work. It's also climate work'. [YouTube Video]; Kelly-Costello, A. (2023, 4 April). 'Where disability and climate meet'. Disability Debrief.
5. Kutz, J. (2022, July). 'Under climate change, care workers often act as first responders'. *Undark Magazine*.
6. Anguiano, D. (2020, 30 December). 'California's wildfire hell: How 2020 became the state's worst ever fire season'. *Guardian*.
7. Ho, Vivian. (2020, 22 August). '"An impossible choice": Farmworkers pick a paycheck over health despite smoke-filled air'. *Guardian*.
8. Wicked Leeks (2023, 27 July). 'Exploits: A new series on people in the supply chain'.
9. Pullman, N. (2023, 30 June). 'We weren't humans: Seasonal workers speak up'. Wicked Leeks.
10. Lynch, J., Cain, M., Frame, D., and Pierrehumbert, R. (2021). 'Agriculture's contribution to climate change and role in mitigation is distinct from predominantly fossil CO_2-emitting sectors'. *Frontiers in Sustainable Food Systems*, 4, 518039.
11. Brewer, G. (2022, 18 August). 'Back to the future: The green revolution'. Kew Gardens.
12. Lawrence, G., and McMichael, P. (2012). 'The question of food security'. *The International Journal of Sociology of Agriculture and Food*, 19(2), 135–142.
13. UN Women (2012). 'Facts and figures: Commission on the status of women'.
14. Navdanya (n.d.). 'Conserving diversity and reclaiming commons'.
15. Atalla, G., and Di Pasquale, G. (2023, 8 November). 'How can workers find their place in the green economy?'. Ernst & Young.

16. UN News (2022, 9 December). '"Just Transition" policies needed to create 20 million green jobs: UN report'.
17. Palladino, L., and Gunn-Wright, R. (2021). 'Care & climate: Understanding the policy intersections'. Feminist Green New Deal Coalition.
18. Kjellström, T. et al. (2019). 'Working on a warmer planet: The impact of heat stress on labour productivity and decent work'. International Labour Organization (ILO).
19. Palladino, L., and Gunn-Wright, R. (2021). 'Care & climate: Understanding the policy intersections'. Feminist Green New Deal Coalition.
20. Palladino, L. M., and Mabud, R. (2021). 'It's time to care: The economic case for investing in a care infrastructure'. Time's Up, Measure Up.
21. Tronto, J. C. (1998). 'An ethic of care'. *Generations: Journal of the American Society on Aging*, 22(3), 15–20.
22. Frida (2018, 28 November). '"Rooted in care: Sustaining movements" volume II of the Young Feminists for Climate Justice Storytelling project'.

32 MAKING KIN WITH THE EARTH

1. Anderson, J. (2015, 30 July). 'How a small-time drug dealer rescued dozens during Katrina'. BuzzFeed News.
2. Burley, S. (2020, 24 August). 'What New Orleans' Common Ground Collective can teach us about surviving crisis together'. Resilience.
3. Dunson, J. (2022). 'Building power while the lights are out: Disasters, mutual aid, and dual power'.
4. The Care Collective (2020). *The Care Manifesto: The Politics of Interdependence* (p. 38).
5. Piepzna-Samarasinha, L. L. (2018). *Care Work: Dreaming Disability Justice*. Arsenal Pulp Press.
6. Ibid.
7. Solnit, R. (2020, 14 May). '"The way we get through this is together": The rise of mutual aid under coronavirus'. *Guardian*.

8. Holloway, B. T., et al. (2023). '"We're all we have": Envisioning the future of mutual aid from queer and trans perspectives'. *The Journal of Sociology & Social Welfare, 50*(1), Article 9.
9. Bergman, C. (2021). 'Mutual aid is kin(etic)'. Grounded Futures.
10. Criales-Unzueta, J. C. (2023). 'From underground subculture to global phenomenon: An oral history of ballroom within mainstream culture'. Vogue; Bailey, M. (2021). 'Structures of kinship in Ballroom culture'. The Architectural Review.
11. Blakemore, E. (2021, 29 January). 'How the Black Panthers' breakfast program both inspired and threatened the government'. History.
12. Ibid.
13. Ibid.
14. Haraway, D. J. (2016). *Staying with the Trouble: Making Kin in the Chthulucene.* Duke University Press
15. Ibid.
16. Latour, B. (2018). *Down to Earth: Politics in the New Climatic Regime.* John Wiley & Sons.
17. Krawec, P. (2022). *Becoming Kin: An Indigenous Call to Unforgetting the Past and Reimagining our Future.* Nation Books.
18. Hayes, K. (2023). '"What kind of story are we setting ourselves up to replicate in the world?" asks "Becoming Kin" author Patty Krawec'. Truthout.
19. Salmón, E. (2000). 'Kincentric ecology: Indigenous perceptions of the human–nature relationship'. *Ecological Applications, 10*(5), 1327–1332.
20. Royal Horticultural Society (n.d.). 'Slime moulds'.

EPILOGUE: ROOTED HOPE: OUR NATURAL CONNECTION

1. ReliefWeb (2015, 24 February). 'Floods in Ghana'.
2. Dii-Osman, R. K. (2022, 22 June). '"Everything is destroyed": Extreme flooding in Ghana tests climate resilience'. The World.
3. Woodland Trust (n.d.). 'Can woods and trees reduce flooding?'.
4. Siddique, A. (2022, 3 June). 'Bangladeshi coastal communities plant mangroves as a shield against cyclones'. Mongabay.

5. UNESCO (n.d.). 'The Sundarbans'.
6. Ghosh, S. (2020, 28 May). 'Erosion, an important cause of mangrove loss in the Sundarbans'. Mongabay.
7. Ibid.
8. Krawec, P. (2022). *Becoming Kin: An Indigenous Call to Unforgetting the Past and Reimagining Our Future.* Nation Books.
9. Ibid.
10. Fisher, M. (2022). *Capitalist Realism: Is there No Alternative?* John Hunt Publishing.

Index

Abanyole community
  (Kenya), 135–6
Abney Park Chapel, Hackney, 235
Abram, David, 94
Accra, Ghana, 196, 199, 290–91
activism, 15–32, 33–40, 53–8,
  59–64, 267–8
  academics and, 181–5
  Black community and, 184–5
  disability and, 265–6
  environmental justice
    activism, 22–32
  rage and 15–64
  religion and, 231–5
  sacrifice and, 264–5
  theory and, 160
Adansi region, Ghana, 99
adobe (building material), 150–52
African Socialism Revisited
  (Nkrumah), 127
African Union, 127
Africanfuturism, 103–4
Afrofuturism, 101–4, 126–7, 134
Agbogbloshie e-waste dump,
  Ghana, 198–9
Aghaji, Daze, 210, 235
agricultural workers, 271–4
Aguon, Julian, 75–6
AI language models, 117–18, 129
'Ain't I A Woman' (hooks), 163
air-conditioning systems, 148–9
Akiyama, Tomohide, 217
Akómoláfé, Báyo, 12, 168, 235–7, 241,
  289, 295
Alabama, United States, 173–4

Albert, Gerrard, 178–80
Albrecht, Glenn, 15, 229
algal blooms, 2
Amazon rainforest, 43–6, 60
Amnesty International, 36
animism, 93–6
Anishinaabe people, 285, 292–3
Ant Studio (design company), 148
Antebellum South, 283
Anthropocene, 256–7
Apple, 113
  Apple Vision Pro, 113, 120
Arias, Kalpana, 119, 121–2
Aristotle, 17
Armstrong, Karen, 210–12
Asaam, Chloe, 197
Atlanta, Georgia, 246–7
Atlas Peak fire, California, 269–70
Atmos magazine, 44, 244
Australia, 85, 115
  Indigenous people, 83

Babangid, Ibrahim, 35
Bacon, Francis, 121
bâdgirs, 149
Bahnson, Fred, 205, 207
Bahuguna, Sunderlal, 50
Balogun, Fehinti, 11, 76, 88
Bangladesh, 291–2
Barad, Karen, 285
Barber, Aja, 196
Barcarena, Brazil, 43
Batammariba people, 151
Bath University, 117
Baubotanik Footbridge, 143

Beachler, Hannah, 132–3
Bean v Southwestern Waste Management case, 171–2
Becoming Kin (Krawec), 285, 292
Benlex Marine Systems, 199
bergman, carla, 283
BERT (language model), 117
Beyonce, 132
Bezos Earth Fund, 116
Bhatt, Chandi Prasad, 49–51
biodiversity conservation, 114, 262–3, 272
biomimicry, 143–4
bird-watching, 224–6
Birdgirl (Craig), 225–6
Black Girls Hike, 226–7
Black Panther (film series), 102, 126–34
   Black Panther: Wakanda Forever, 126–34
Black Panthers, 280, 283–4
   Breakfast Programme 283–4
Black Radical Tradition, 184–5, 188, 190
'Black to the Future: Interviews with Samuel A. Delany, Greg Tate and Tricia Rose' (Dery), 101
Black women, 17–18, 129
   feminist movement, 163
   technology, 128–9
Black2Nature (charity), 226
Bleeding Tree, The (Starling), 243
Boeri, Stefano, 144
Boka-Batesa, Destiny, 22–4
Bolivia, 264
Bolsanaro, Jair, 45
Botswana, 152–4
Bowman, Jamaal, 275
BP (British Petroleum), 34, 261
Braiding Sweetgrass (Kimmerer), 61, 87, 95
Brauer-Maxaiea, Nyeleti, 22–4
Brazil, 10, 40, 41–7, 264
   gold rush, 41–2
Brexit, 272
Britain *see* United Kingdom
brown, adrienne maree, 21, 58

Brown, Mack, 245–6
Brown, Rev. Luther G., 27
Brunel University of London, 134
building materials, 150–52
Bullard, Linda, 170–73
Bullard, Robert, 170–73, 180
Buolamwini, Joy, 129
Burkina Faso, 151–2
Burnett, Jasmine, 21
Burwell, Dollie, 25–6
Butler, Octavia E., 76–7, 104–7

Cahto people, 82
Cailleach (ancient Celt), 86
California, United States, 83–4, 219, 269–71
'Callings & Roles for Collective Liberation' framework, 57
Cambridge Centre for Carbon Credits (4C), 116
Cambridge Climate Justice, 181
Cambridge Stop the War Coalition, 181
Cambridge Student Action for Refugees, 181
Cambridge University, 6, 116, 142
Campbell-Stephens, Rosemary, 10
Can I Live? (Balogun), 76
Canada, 96
Canavan, Gerry, 107
'Cancer Alley' (Louisiana), 171, 187–90
Cape Coast Castle, Ghana, 99
capitalism, 3, 77–81, 84–9, 108, 192, 234, 254, 256, 293
Capitalist Realism (Fisher), 293
carbon capture and storage, 114, 116
carbon credits, 116, 118, 236
carbon emissions, 114–15, 151, 195, 211, 264, 274–5
'carbon footprint', 261
carbon offsets, 118, 259, 263
care, 9, 11–12, 251–98
   care infrastructures, 274–7
   climate care, 259–69
   collective care, 251–5, 257, 280
   emergency-response care, 278–82

environmental care systems, 269–74
self-care, 254–7
social care, 253–4
'Care & Climate: Understanding the Policy Intersections' report (Palladino & Gunn-Wright), 276
Care Collective, 280
Care Manifesto, The (Care Collective), 258
Care Work (Piepzna-Samarasinha), 265
care workers, 270–71, 275–7
Carson, Rachel, 159
Carver, George Washington, 212
'Case for Climate Rage, The' (Westervelt), 16–17
Celidwen, Yuria, 73, 241–2
Celts, 86, 244
Central Park birdwatching incident (New York), 224–5
Chamoli district, Uttar Pradesh, 10
ChatGPT, 117
Chávez, César, 28
Chavis, Ben, 27–8
Chevron, 35
Chieza, Natsai Audrey, 144
China, 94, 219
Chipko Movement, India, 48–53
Choked Up (charity), 22–4
Christchurch, New Zealand, 60
Christianity, 206, 231–3
Chrysalis Youth Fund, 244
Church Forests, Ethiopia, 205–9
circular economy, 193, 195, 197
Civil Rights Act (US, 1964), 173
clay-based technology, 148–52
Clean Air Act (UK), 23
'Climate Action Venn Diagram' (Johnson), 57
'Climate Anxiety: An Illness of the System' (Siddiqa), 229
climate crisis, 17–18
 art and culture and, 74–6, 91–3
 climate grief, 238–47
 colonialism and, 6–8
 environmental activism and, 15–32, 53–8
 racial discrimination and, 171–3
 spirituality and, 210–12
 technology and, 113–25
climate disaster, 290–95
ClimateInColour, 6–7
 Reads, 165–6
 'The Colonial History of Climate', 6–7
Clinton, Bill, 29
Clyde River, Scotland, 191
Coastal Environments Inc (CEI), 187–8
collective care, 251–5, 257, 280
collective imagination, 67–77, 87, 108
collective memory, 81
collective wisdom, 9, 165, 174, 175–80
Colombia, 264
colonialism, 3, 6–8, 33, 77–80, 84–9, 102, 104, 108, 120–21, 177, 231–3, 262
coltan (mineral), 130
Combahee River Collective, 129
Common Ground Collective, 279–80
Community Remembrance project, 246
compost, 288–9
Conness, John, 83
Conservative Party (UK), 253
Consumed (Barber), 196
'CoolAnt' air-conditioning system, 148–9, 154
Cooper, Amy, 224–5
Cooper, Christian, 224–5
Cornwall, England, 242
council estates, 280–81
Covid-19 pandemic, 220–21, 270–71
Craig, Mya-Rose (BirdGirl), 225
Crenshaw, Kimberlé, 18
Crutzen, Paul, 256
cultural burning, 83–5
Cyclone Yaas, 292

INDEX

Damh the Bard, 86
Dasholi Gram Swarajya Mandal (DGSM), 49–50
Davis, Angela, 18, 166, 182–3
Dawn of Everything, The (Graeber & Wengrow), 84
Dead White Man's Clothes research project, 197
Death Alley *see* Cancer Alley (Louisiana)
decolonisation, 79–81
Defebaugh, Willow, 244–5
degrowth theory, 193–4
Democratic Republic of Congo, 118, 130, 255
Dery, Mark, 101–2
Descartes, René, 233
Design for Planet (COP26 event), 191–2
Devi, Gaura, 51–2
Dickinson, Emily, 298
'Discourse and Method' (Descartes), 233
Dochartaigh, Kerri ní, 286–7
Dooh, Barizaa, 37–8
Dooh, Eric, 38
Dordor, Naa Asheley, 198
Doughnut Economics (Raworth), 192–3
Douglass, Frederick, 98
dreams, 87–8
Drive Your Plow Over the Bones of the Dead (Tokarczuk), 59
drought prediction, 135–7
druids, 86
Dumping Dixie: Race, Class and Environmental Quality (Bullard), 172

Earth (planet), 103
Earth Church, 234–5
Earth Day, 33, 234
Earth Emotions (Albrecht), 15
eco-anger, 15–16
eco-anxiety, 218, 229
ecofascism, 61, 72, 263
'ecology of perception', 95

economics, 191–4, 252–4
circular economy, 193, 195, 197
degrowth theory, 193–4
Doughnut Economics, 192–3
Ecuador, 60
Efunwale, Obatala, 203, 211, 231
Egypt, 115, 149
El Paso, Texas, 60
electric vehicles, 78–9, 116
Eliasson, Olafur, 76, 90–93
Ellen MacArthur Foundation, 193
Elmina Castle, Ghana, 99
Embobut Forest, Kenya, 263
Emelle, Alabama, 171
Emergence Magazine, 86, 94, 205, 286
employment, 274–7
Entangled Life: How Fungi Make Our Worlds, Change Our Minds, and Shape Our Futures (Sheldrake), 73–4
environmental justice activism, 22–32
environmentalism, 5, 7–8, 10, 85, 209, 233
Equal Justice Initiative, 246
Equiano, Olaudah, 237
Eshete, Alemayehu Wassie, 206–8
Ethiopia, 205–9
Church Forests, 205–9
Ethiopian Orthodox Tewahedo Church, 205–9
Evaristo, Bernardine, 18
Evon, Kim, 270
Extinction Rebellion, 18, 21

Faber Futures (biodesign lab), 144
Facebook, 126
faith *see* religion
fashion waste and pollution, 144, 161, 194–8
Fatinikun, Rhiane 226
Feminist Green New Deal, 276
Ferruccio, Ken, 27
Ficus elastica (Indian rubber tree), 139

First National People of Color Environmental Leadership Summit (1991), 172
Fisher, Mark, 293
Flock Together, Essex, 226
floods, 49–50, 290–92
Flowers, Catherine Coleman, 173
Floyd, George, 79, 224
For the Wild podcast, 241
Forensic Architecture (research group), 10–11, 188–90
forest fires, 83, 269, 279
'Forest Satyagraha' protest, India, 48–9
Forest Therapy, 217–18
Formosa Plastics Group, 187–9
Fossil Free London, 39
fossil fuels, 33–40, 60, 181–2, 274
   fossil fuel industry, 261, 267–8
'Four Roles of Activism' (Moyer), 55–7
Fox, Keolu, 286
Franco, Marielle, 45
Free African Society, 282
Freire, Paulo, 164
Friends of the Earth, 34, 37–8
Fruity Walks, 227
Futures of Black Radicalism (Johnson & Lubin eds), 184

Ga tribe, Ghana, 243
Gandhi, Mahatma, 49
gardening, 216
Gawi Wachi, Mexico, 286
Gay, Roxane, 18
Gaza Strip, 100
Gebru, Timnit, 129
Ghana, 5, 127, 130, 139, 194, 196–200, 215, 218
   floods, 290–92
   grief, 238–40, 243–4
   Kantamanto Market, 161, 196–8
   Sankofa, 123, 175
   slavery and, 98–9
Ghosh, Amitav, 72
giant sequoia, 70, 82–3
Gibson, Jabbar, 279–80

Gilligan, Carol, 252
glacier melt series 1999/2019, The (Eliasson), 92–3
glaciers, 91–4
Global Innovation Hub, 115
Global Majority, 10
global warming, 147–8, 219
Global North, 18, 61, 72, 79, 94, 161, 190, 229, 257
Global South, 30, 161, 190, 229, 231, 255, 268–9, 272
   definition of, 5–6
Global Witness (international NGO), 45
Goi, Ogoniland, Nigeria, 37
Gold Coast, 127
gold mining, 130, 239–40
Google, 117, 126, 129
Gordon, Christopher, 291
Gorelick, Dan, 234
Graeber, David, 84
'Great Derangement', 72–3
green hydrogen, 115
Green Revolution, 272
Greenpeace, 35
grief, 238–48, 293
Grief Is a Thing with Feathers (Porter), 243
Grizedale Forest, Lake District, 221–2
Groundwork (charity), 220
Grzimek, Bernhard, 263
Guardian, 15, 33, 46, 271
Guha, Ramachandra, 52
Gunn-Wright, Rhiana, 276
Gurunsi people, 151
Guterres, António, 16

Halstead, John, 103
Haraway, Donna, 284–5, 289
Hargreaves, Samantha, 103
Hariramani, Divya, 227
Hartman, Saidiya, 78
Hawai'i, 286
healing, 9, 11–12, 205–48
   natural world and, 216–27
Held, Virginia, 252

INDEX

Hernandez, Isaias, 166–7
Hersey, Tricia, 246
Hickel, Jason, 193–4
HIV/AIDS, 283
Holiday, Ryan, 182
Holloway, Zena, 144–5
hooks, bell, 160, 163–4, 168, 183–5, 228
Hoover, J. Edgar, 284
Hope in the Dark (Solnit), 63, 119
Houston, Texas, 170–71
'how the wonder of nature can inspire social justice activism' (brown), 21
Hudson, A. J., 100
Hughes, Seth, 242–3
Hull University, 117
Hurricane Katrina, 278–80

Iceland, 91
Imafidon, Anne-Marie, 128
imagination, 9, 11, 67–109
    collective imagination, 67–77, 87, 108
In real life exhibition (Eliasson), 93
India, 6, 48–53, 147–8, 273–4
Indigenous Innovation Summit (150th, Winnipeg), 122
Indigenous peoples, 82–7, 121, 228–33, 285–6
    animism, 93–4
    biological knowledge, 273–4
    conservation, 263–5
    cultural burning, 83–5
    environmentalism and, 82–5, 97
    innovation/technology and, 120–23, 134–7, 140–43
    languages, 80
individualism, 255–6, 281, 284
Industrial Revolution, 120
innovation, 9, 11, 109–10, 113–56
    Indigenous peoples and, 120–23, 134–7, 140–43
    rooted innovation, 123–5, 154–5
Innovation Centre for Applied Sustainable Technologies (iCAST), 117

Innu people, 96
Instagram, 223
Institute for Advanced Architecture of Catalonia (IAAC), 152–3
International Labour Organisation, 275
'Intersectional Experiences of Black Women in Computing, The' (Rankin & Thomas), 129
Interstellar (film), 102–3
Iran, 149
Ireland, 86
Is the River Alive? (Macfarlane), 95
IT1KI (Information Technology and Indigenous Knowledge with Intelligence), 136–7
iwígara, 286
Iyer, Deepa, 57

Jamie, Kathleen, 191
Japan, 217
    Shinto, 217–18
Johnson, Ayana Elizabeth, 57
Johnson, Gaye Theresa, 184
Jones, Lucy F., 86, 216–17, 287–9
Just Stop Oil, 18

Kalahari Desert, 152–3
Kānaka Maoli (Hawai'ians), 286
Kantamanto Market, Ghana, 161, 196–8
Kapok tree (*Ceiba pentandra*), 70
Kaur, Jasbir, 52
kayayei women, 197–8
Kennedy, Ryan, 131
Kennerley, Rachel, 38
Kenya, 135–7, 234, 263
Kéré, Francis Diébédo, 152
Khan, Sadiq, 23
Khasi people, 140–41
Khoikhoi people, 152
Khongthaw, Morningstar, 141–2
Kimmerer, Robin Wall, 61, 87, 95
'Kincentric Ecology: Indigenous Perceptions of the Human-Nature Relationship' (Salmón), 286

INDEX

King Jnr, Martin Luther, 246
Kipling, Rudyard, 231
Kissi-Debrah, Ella, 22
Klein, Naomi, 241
kola nut, 175
Krawec, Patty, 3, 80, 285, 292–3
Kropotkin, Peter, 283

Labalme, Jenny, 29
Lake Chad, 31
Lake District, UK, 221–2
Lake Tana, Ethiopia, 207
Lanham, J. Drew, 62, 224
Lavigne, Sharon, 186–7, 189–90
Le Guin, Ursula K., 94
Lead to Life collective, 247
Lee, Jessica J., 287
Lee, Sam, 86
legal personhood, 178–80
Lemonade (Beyoncé), 132
Less Is More: How Degrowth Will Save the World (Hickel), 194
Levin, Kelly, 116
Levy-Rapoport, Noga, 15, 54
Lewis, Sally, 270
Lipsitz, George, 184, 190
lithium mining, 78, 118
Live Frankly (media platform), 272
living root bridges, 139–42
Living Root Foundation, 141
Lloyds Bank, 117
Lo-TEK (Watson), 141
Loach, Mikaela, 39
London, UK, 22–3
'Long Memory, The' (Phillips), 81
Lorde, Audre, 18, 20, 182–3, 254
Losing Eden (Jones), 216–17
Loss and Damage Fund, 115–16
Lowe, Miranda, 12, 162–3
Lowman, Margaret, 207
Lowndes County, Alabama, 173–4
Lubin, Alex, 184
Ludwig, Ferdinand, 143

Mabud, Rakeen, 276
MacArthur, Ellen, 193
Macfarlane, Robert, 11, 94–6

Madhya Pradesh, India, 48–9
Magnason, Andri Snær, 92
Malpas, Imogen, 234
Mammoth Tree (giant sequoia), 82–3
Mantawoman, 234
Māori people, 176–9
Mapondera, Margaret, 103
Maroons, 42
Marshmallow Laser Feast (MLF), 70–71, 113
Masinde, Muthoni, 11, 134–8
mass extinctions, 1–2
Material Cultures (design and research organisation), 150
Matrescence (Jones), 287
May, Katherine, 12, 257, 264–5, 277, 280
Mbeere community (Kenya), 135–6
Mbewe, Masiyaleti, 102
Me-Wuk Nation, 82–3
media, 16, 19, 31, 55
Meghalaya, India, 139–43
Mehta, Jojo, 59
mental health, 216–17
de Mestral, George, 143
meteorology, 135–7
Mexico, 246
Miles, Ellen, 225
mining companies, 118
Ministry for the Future, The (Robinson), 147
misogyny, 17
Mississippi River, 186–90
Montana, United States, 60
'Most Dangerous Story Ever Told: Ecological Collapse, Progress, and Human Destiny, The' (Halstead), 103
Mother Tree (giant sequoia), 67–9, 82–3
motherhood, 287–8
Movement for the Survival of Ogoni People (MOSOP), 35–6
Moyer, Bill, 55–7
Mozambique, 137
Musk, Elon, 103

Muteshekau-Shipu River, Quebec, 96
mutual aid, 282–4
My Bondage and My Freedom (Douglass), 98
Myers, Natasha, 90, 285

Naidoo, Kumi, 75
Nakate, Vanessa, 234
Napa Valley, California, 269–70
natural disasters, 281–2
Natural History Museum, London, 12
natural world
  disconnection with, 228–33
  healing and, 216–27
  racial prejudice and, 224–5
Nature Is a Human Right, 54
nature technology, 116–17
Navdanya (India), 273–4
New Ghanzi project (Botswana), 152–4
New Orleans, Louisiana, 279
New Zealand, 96–7, 176–9
NHS (National Health Service), 253
Nigeria, 10, 32, 33–40
Nigerian Petroleum Company, 35
Nkrumah, Kwame, 127
No Country for Eight-Spot Butterflies (Aguon), 75
Noddings, Nel, 252
Nolan, Christopher, 102
Normalized Difference Vegetation Index, 220–21
Norsk Hydro, 43–4
Northeast Community Action Group, 170–73
Norway, 43–4, 115
Not Just You (Tsui), 218
Nowadays on Earth, 119

Octavia's Brood: Science Fiction Stories from Social Justice Movements, 81
Ogilvy (advertising firm), 261
Ogoni 9 (Nigeria), 10, 32, 33–40

Ogoniland, Nigeria, 33–8
oil spills, 33–8
Ok (Okjökull) glacier, Iceland, 92
Okorafor, Nnedi, 103
Oladosu, Adenike, 11, 30–31
Olanipekun, Ollie, 226
Olympic Games, 2016 (Rio de Janeiro), 43
'Open Letter to the Philanthropic Community: Harnessing the Power of Arts and Culture for Humanity's Survival' (Naidoo), 75
oppressive systems, 163–4, 184, 190, 245, 264
Or Foundation, The, 161, 194–200
  No More Fast Fashion Lab, 194, 197
Organisation for African Unity, 127
Osuala, Uzoma Samuel, 140
Oteng, Sammy, 197
Our Time on Earth exhibition, 69–70
'Out of Bounds: Equity in Access to Urban Nature' (Groundwork), 220
Oxford University, 117

Pakistan, 149
Palas por Pistolas project (Mexico), 246
Palestine, 100
Palladino, Lenore, 276
Pan-Africanism, 127
Pará, Brazil, 43–5
Parable of the Sower (Butler), 104–7
Paradise Local Nature Reserve, Cambridge, 214
Paris Climate Agreements, 39
Parks, Janile, 188
Pedagogy of the Oppressed (Freire), 164
Peehee Mu'huh, Nevada, 78–9
Perera, Nadeem, 226
petrochemical industry, 187–90
Phillips, Morrigan, 81

INDEX 339

Piepzna-Samarasinha, Leah Lakshmi, 251, 265–6, 281–2
'planthropscene', 90
Pleasure Activism (brown), 58, 63–4
Png, Marie-Therese, 234
pollution, 161–2, 187–90
  air pollution, 22–3, 148, 166, 266
  fashion waste and pollution, 144, 161, 194–8
  water pollution, 30, 37, 44–5, 96–7
polychlorinated biphenyls (PCBs), 24–6
Pomo people, 82
Porter, Max, 243
Principles of Environmental Justice, 172
Principles of Working Together, 172
public health, 22, 173–4, 216–19

Qatar, 149
Quebec, Canada, 96
quilombo resisters (Brazil), 10
quilombola (Afro-Brazilian community), 41–7

racism, 17–18, 224–5, 245–6
  racial segregation, 23–4
  racial violence, 246–8
rage, 9–11, 15–64, 108, 240–41
  activism and 15–64
  masculine anger, 17
Rahim, Malik, 280
Rainham Marshes, Essex, 226
Raman-Middleton, Anjali, 22–4
Rankin, Yolanda, 129
Raworth, Kate, 191–4, 197
RawTools, 246
'Rebel with a Cause: How to Become an Activist' (Levy-Rapoport), 54
Reddy, Jini, 222
Reddy, Trusha, 103
redlining, 23–4
Regan, Michael S., 29
regenerative design, 143
religion, 206, 208–12, 218, 231–6
renewable energy, 114, 116, 234

Revolution Will Not Be Litigated: How Movements and Law Can Work Together to Win, The 170
Ribeira Valley, Brazil, 41–3
Ricketts, Liz, 197
Rio Tinto, 118
RISE St James community group, 186–9
Riverford (organic farm), 272
Robinson, Cedric J., 184–5
Robinson, Kim Stanley, 147
Rome, Italy, 219
rooted innovation, 123–5, 154–5
roots, 1–2, 144–6, 291–2
  in the living world, 3–5
  living root bridges, 139–42
  mass extinctions and, 1–2
Roswell, Georgia, 244
Roy, Arundhati, 103
Royal Institute of British Architects (RIBA), 143
Royal Museums Greenwich, London, 237
Rules for Radicals (Holiday), 182

Sacred Nature (Armstrong), 210
Salmón, Enrique, 11, 286
San community (Botswana), 152–4
Sanbao, China, 219
Sankofa (Ghana), 123, 175
Santos, Teresa, 270
Saro-Wiwa, Ken, 10, 34–7
Schumacher, E. F., 155
science, 159
science fiction, 101–3, 106
SciFest Africa 2021, 137
Scotland, 86, 191
Seifert, Jeremy, 205
sequoia, giant, 70, 82–3
Sequoia National Park, 83
Serbia, 118
Service Employees International Union, 270
Sheldrake, Merlin, 73–4
Shell, 10, 33–9, 187
Shinrin-Yoku, 217
Shiva, Vandana, 48, 273

'Should Trees Have Standing' (Stone), 178
Siddiqa, Ayisha, 229
Sierra National Forest, 83
Simon Company (sporting goods manufacturer), 50
Sinclair, Murray, 122–3
Sinkyone people, 82
Siripurapu, Monish, 148–9
Sitka spruce, 221–2
Skinner, Branson, 197
slave trade, 41–3, 98–100, 104, 120, 187, 237
Slow Factory, The, 57
Small Is Beautiful: Economics As If People Mattered (Schumacher), 155
'Social Change Ecosystem Map' (Iyer), 57
social media, 55, 59, 234, 294
'Society, Technology and Cosmology in *Black Panther*' (Kennedy), 131
Socorro Silva, Maria do, 44–6
soil, 288–9
  compost, 288–9
  microbes, 216
solastalgia, 229–30
Solnit, Rebecca, 12, 63, 119, 282, 294–5
South Africa, 49, 137
Sparrow Envy: Field Guide to Birds and Lesser Beasts (Lanham), 224
spirituality, 210–12, 234–6, 242
  see also religion
St James Parish, Louisiana, 186–90
Starling, Hollie. 244
Staying with the Trouble (Haraway), 284–5
Steer, Andrew, 116
Stefanovic, Sofija, 118
Stemettes, 128
Stoermer, Eugene, 256
Stone, Christopher, 178
Stop Ecocide International, 59, 82
storytelling, 74, 100–101, 104, 141

student movement, 181–4
Sudan, 100
'swords to ploughshares', 246

Tate Modern, London, 93
Te Awa Tupua Act (New Zealand, 2017), 179
technology, 113–56
  climate crisis and, 113–25
  Indigenous peoples and, 120–23, 134–7, 140–43
  rooted innovation, 123–5, 154–5
terrafurie, 15
Tetteh, Yvette, 199–200
Texas, United States, 170
theory, 9, 11–12, 159–201
  children and, 168
  definition of, 160–61
  economic theories, 191–4
'Theory as a Liberatory Practice' (hooks), 163
Thomas, Jakita, 129
3D printing, 152, 154
Thunberg, Greta, 31
Togo, 151
Tokarczuk, Olga, 59
'Touching the Earth' (hooks), 228
'Toxic Waste and Race in the United States Report' (United Church of Christ), 28–9
toxic waste, 24–9
tree therapy, 90–91
'Treehugger: Wawona' project, 70
'treehuggers', 48, 50
Tronto, Joan, 277
Trussell Trust, 253
Tsui, Tori, 218

UN (United Nations), 76, 114–15, 263
  Sub-Commission on the Promotion and Protection of Human Rights, 35
UNFCCC (United Nations Framework Convention on Climate Change)

COP26 (Glasgow, 2021), 76, 191, 267–8
COP27 (Sharm el Sheikh, 2022), 114–16
COP28 (Dubai, 2023), 267–8
United Church of Christ, 28
United Farm Workers of America union, 28
United Kingdom, 23, 85–6, 115, 151, 196, 225
  agricultural workers, 272
  ancient Indigenous people, 85–6
  care work, 253
  climate research investment, 117
  colonial past, 79, 127
  Industrial Revolution, 120
  poverty, 253–4
  renewable energy, 234
  social support, 253–4
United States, 24–32, 178, 186–90
  Black communities, 171–3, 245–7
  Environmental Protection Agency, 24, 29
  Native Americans, 82–5, 232
  Office of Environmental Justice and External Civil Rights, 29
  renewable energy, 234
  wildfires, 269–71
urban landscapes, 219–21
Uttar Pradesh, India, 10, 48, 52

V&A Dundee, 191
van der Hout, Floor, 180
Vaughan-Lee, Emmanuel, 210–11
velez, brontë, 11, 247
'Vertical Forest', Milan, 144
Volta River, Ghana, 199–200

Wagstaff, Dulcie, 216
Wakanda (fictional African country), 126–34
Wanderland (Reddy), 222
Ward Transformer Company, 24
Warren County, North Carolina, 24–30, 161
'Waste Age' Exhibition, Design Museum, London, 150–51
waste disposal sites, 170–71
Watson, Julia, 11, 141, 145
We The People Nigeria (community group), 39
Wellcome Photography Prize, 216
Wengrow, David, 84
Westervelt, Amy, 16–17
Whanganui Iwi people, 177–9
Whanganui River, Northern Aotearoa (New Zealand), 96–7
wheatgrass seed, 144–5
'When You Could Hear the Trees' (Dochartaigh)
'White Man's Burden, The' (Kipling), 230–31
White, Rev. Leon, 27
Wicked Leeks, 272
wildfires, 269–71
Woman Who Does Not Fear, The (catamaran), 199–200

Yazd, Iran, 149
Yellop, Alexandra, 234
yew trees, 244
Yosemite Valley, California, 83
Young-Lutunatabua, Thelma 282
Yu-Pearson, Hannah, 234